OPTICAL SIGNALS

Animal Communication and Light

OPTICAL SIGNALS

Animal Communication and Light

Jack P. Hailman

Department of Zoology
University of Wisconsin

INDIANA UNIVERSITY PRESS

Bloomington & London

Published in Canada by Fitzhenry & Whiteside Limited, Don Mills, Ontario

Manufactured in the United States of America

Library of Congress Cataloging in Publication Data

Hailman, Jack Parker, 1936–
    Optical signals.

    Bibliography: p.
    Includes index.
    1. Animal communication. 2. Color of animals.
I. Title.
QL776.H34   591.5'9   76-48501
ISBN 0-253-34254-6

FOR LIZ,

accomplice, belle, cook, diving partner, Eric's mother, family bond, girl-
friend, helpmeet, inspiration, Jack's
purpose, Karl's mother, lover, maid,
naturalist, oenologist, programmer,
queen, runner, scholar, tranquilizer,
umpire, valentine, wife, XXX, youth,
zoologist.

CONTENTS

LIST OF FIGURES

LIST OF TABLES

LIST OF EQUATIONS*

*All variables appear in the Index to Subjects.

PREFACE and ACKNOWLEDGMENTS

This monograph results from my inability to follow instructions. Professor Thomas A. Sebeok kindly asked me to provide the chapter on optical communication for his new volume *How Animals Communicate* (Sebeok, 1977), but my manuscript exceeded the assignment in both length and scope. Indiana University Press then invited me to extract the empirical review for Sebeok's book (see Hailman, 1977a) and formulate the more wide-ranging material into a monograph devoted to optical signals. Hence the book in hand.

I have tried to strike a balance between a self-contained monograph and one that did not repeat the substance of my chapter in Sebeok's book. The compromise results in two factors that readers who are ethologists may find peculiar: some explanations of quite elementary ethological notions are contained herein, yet some traditional and important topics are treated only briefly. The first factor results from the attempt to make the monograph intelligible to readers not trained in ethology, and the second from the attempt to avoid overlap with the chapter, to be published nearly simultaneously. Frequent reference to the chapter helps preserve the continuity.

This whole endeavor would not have been possible without the aid and abetting of many individuals, who in no way bear responsibility for any outrages contained herein. The prime mover was Peter Marler, who suggested that I write the chapter for *How Animals Communicate*, thus succeeding his forerunner in Sebeok's (1968) earlier *Animal Communication*. Thomas Sebeok himself was a constant source of support for both the chapter and this monograph that grew from it, and has kindly contributed some words in the fore. John Gallman, Director of Indiana University Press, from his initial invitation to the last throes of production, kept close touch with general progress, while John Vint and particularly Natalie Wrubel of the Press struggled with the details.

Those who contributed to the manuscript, directly or indirectly, were many. Not only those whose works are specifically cited, but many others whose writings have influenced my thinking, share in the thanks. Some of my ideas were discussed specifically with Wolfgang M. Schleidt and Gordon R. Stephenson; my colleagues Jeffrey R. Baylis and Warren P. Porter were particularly useful sounding boards, who also made special efforts to provide examples and data to secure certain points. Anywhere from pieces to substantial chunks of the manuscript were read by Lee McGeorge, Daniel Rubenstein, Robert G. Jaeger, Scott R. Robinson, Peter H. Klopfer and Alan H. Brush. I was fortunate enough to have no fewer than five persons who suffered through the entire manuscript at one stage or another: to Jeff Baylis, previously mentioned, Steve Witkin, Lincoln Fairchild, Edward H. Burtt, Jr. and Timothy D. Johnston both I and the readers are grateful, whether the latter realize it or not.

Technical production also involved many persons to whom I owe thanks. The figures were all executed by Cheryle Hughes, who translated figures from the literature, my scrawled graphs, my photographs and even my field sketches into clear, consistently styled and charming illustrations. The photographic preparation of figures for publication was competently handled by Claudia Lipke, under the watchful eye of Don Chandler. The important and often overlooked task of indexing was performed by Scott and Karen Robinson, and by my wife, Liz, who also proofread the entire text and helped in ways too numerous to document. Indeed my entire family deserves praise for surviving nearly a year of weekends in which I spent almost every waking moment at the typewriter instead of doing something productive. Despite the pressures of time, Jay Parker, who typed the camera-ready manuscript for photo-offset, might have done a better job.

J.P.H.

Village of Shorewood Hills
March 1977

By Thomas A. Sebeok

In an early approximation to the typology of visual
vs. auditory signs--the sole dichotomy considered in his
1963 semiotic analysis--Roman Jakobson advanced the thesis that vision is man's paramount spatial sense and audition his foremost temporal one. While, in the former,
the principal structuring device is simultaneity, in the
latter it is successivity. He spelled out why, in oppo-
sition to the synchronous semiotic type, vision, the sequential type, audition, requires a compulsory hierarch-
ial buildup from discrete elementary components organized
to serve a given communicative purpose. Jakobson's concern was with man alone, but--as I had occasion to point
out in a typologically oriented supplementary sketch on
chemical signs written especially for his 70th birthday--
the contrasts he developed seemed so neat chiefly because
they excluded from consideration all of the other multif-
arious media of communication at the disposal of the vast
gamut of organisms, including man. It now turns out, however, that a popular model of the bicameral mind, with two
underlying cerebral hemispheres ostensibly specialized to
perform quite different but complementary tasks, lends added credence to Jakobson's view, as is palpable especially
from the line of research pursued by Gordon H. Bower.

A few years before Jakobson's analysis, in 1960,
Frank A. Geldard, the premier human sensory physiologist,
following upon the summary work of R.H. Henneman and E.R.
Long in the mid-1950's, reviewed the conditions under
which visual coding seems to be indicated for man, contrasting this mode for casting messages with auditory coding. He identified six such requirements. In reviving
"all this debate about eyes and ears," Geldard's motiva-
ting purpose was to raise his voice about still other possibilities, in particular, the tactile channel, and to focus on the human integument housing several modalities,
"rivaling the ear as a temporal discriminator...and shar-
ing with the retina the property of somewhat orderly spatial extension... ."

By 1964, it appeared pressing to me that we learn to
predict the pivotal characteristics of every kind of sign
propagated in no matter what form of physical energy. Thus
I attempted to compare, in summary fashion, the advantages

and disadvantages of chemical, optical, tactile, and a-
coustic systems--merely four among a multiplicity of
other channels--gently probing why signs have the characteristics they do, but making it clear that "at this
stage" the problem of typology could be adumbrated only
in very modest, preliminary ways, for the sources of our
ignorance were manifold. The existence and functioning
of the aforementioned systems, and a few others--such as
the increasingly appreciated electric channel, the semiotic properties of which have ingeniously been construed
by Carl D. Hopkins--are certain, but, as the great animal
psychologist, Heini Hediger, has repeatedly underlined,
we consistently tend to underestimate the speechless crea-
tures with respect to their sensorial efficiency. He
points, for example, to the literature on the African rhinoceros, still replete with references to its extraordinary short-sightedness. Investigations carried out in the
Zurich Zoological Garden during his directorship have, in
fact, proved that the black rhinoceros has a better eye-
sight than the elephant and many species of antelope. The
notorious Clever Hans episode, the far-reaching semiotic
dimensions and profound lessons of which have, as yet,
been scarcely understood, crucially hinged upon Oskar
Pfungst's demonstration that the horse--a paradigmatic
ancestor of the Clever Dolphins of yesteryear and the
Clever Apes, from the chimpanzee Washoe to the gorilla
Koko, of our decade--is capable of perceiving the raising
of eyebrows, the dilation of nostrils, or even uninten-
tional movements in the human face of less than one-fifth
of a millimeter. We now know that "muscle readers"--like
the amazing Eugen de Rubini--"clairvoyants," "rod diviners,"
and "psychics" or "seers" of the same ilk, can exceptionally do almost as well as "talking" horses, dogs, pigs,
goats, geese, or, most recently, a Greater Spotted Woodpecker, whose linguistic performance has already been de-
clared analogous to Washoe's and Sarah's, in decoding wit-
tingly or unwittingly emitted, extremely subtle visual
cues, some wide ranging implications of which for both
science and daily conduct have been explored with insight
and imagination by Robert Rosenthal. In a 1612 manual
dealing with *The Art of Juggling or Legerdemain*, the first
English-language volume devoted to conjuring, Samuel Rid
commented, with shrewd percipience, on Morocco, the white
steed known to Shakespeare as "the dancing horse," and
which was capable of stamping out with its hoof the total

number of spots on a pair of rolled dice, the number of
shillings in a person's pocket, or the age of anyone who
whispered it to his exhibitor: "mark the eye of the horse
is always upon his master [the Scotsman John Banks, who,
along with his learned horse, was, in the end, allegedly
burned for witchcraft in Rome], and as his master moves,
so goes he or stands still, as he is brought to it at the
first... . And note also that nothing can be done but
his master must first know, and that his master knowing,
the horse is ruled by signs. This if you mark at any time
you shall plainly perceive."

    As von Uexküll has so vividly impressed upon us, the
Umwelt, or phenomenal world, of each animal is shaped (and
sometimes colored) by the very particular configuration of
its sensorial apparatus. And as he had further observed,
many of these worlds are necessarily invisible to humans.
Thus we are incapable of perceiving ultraviolet, bordering
on the x-ray region to about 400 nm, which is readily distinguishable by the honey bee, as Karl von Frisch has demonstrated, abridging the visible spectrum in the red but
extending it to shorter wavelengths. Men likewise cannot
see in the dark without special enhancing instrumentation,
but infrared sensitive nocturnal mammals may come equipped,
so to speak, with their own accouterment, the iridescent
pigment choroid coat known as the tapetum lucidum, enabling
them to exploit the faintest sources of ambient light.

    Not only are the respective ranges of the sensorium
of animals more or less at variance with ours and, of
course, with those of one with another, yielding, at best,
overlapping *Umwelten,* but also their relative acuity is
sure to differ considerably. For instance, the sharp eye
of an African vulture is capable of discriminating from a
height of 4,000 m whether an object lying on the ground is
a sleeping gazelle or the corpse of one. Hediger justly
points out that "this is all the more extraordinary when
we consider that, even with field binoculars, we are unable
to identify the bird at such a height."

    I was fascinated recently to learn that the recipro-
cal exchange of visual signs between an animal sender and
a receiver of another species may not only be imperceptible
to a human observer--one would expect reaction times to be
often much faster than ours, as elegantly demonstrated,
since the early 1960's, by W.H. Thorpe in his acoustic
work with several species of *Laniarius* antiphonal duet
patterns ("the timing of the singing is normally so per-

fect that in no case would one suppose that two birds were involved...")--but that some such signs might be actually reduced to zero. A zero sign is an important semiotic category, not at all the same as the mere absence of communication. The evidence for this comes from the Dutch psychologist F.J.J. Buytendijk's slow-motion analysis of a film depicting a combat between a mongoose and a cobra, where the reaction time between the movements of both predator and snake were (in certain parts) zero. This means that such messages are not actualized but merely anticipated, much as they may be in a well-rehearsed *pas de deux* in a ballet, or, as Hediger surmises, in circus acts, "when the trainer causes the panther sitting on the pedestal to strike out with a fore-paw and draws back in precise accord with the movement, or when the springboard acrobat adjusts his reaction to the blow of the elephant's foot at the other end of the plank." D.N. Lee's observation that body movements, *e.g.*, sway, with respect to the layout of environmental surfaces can be driven phasically by minimal oscillations in optical expansion and contraction, or flow patterns at the eye, thus implies not only a mathematical correlation between coordinative structures and perceptual variables but that this relationship can be specified in highly precise fashion.

For the correct evaluation of the myriad events that take place between animal and animal or man and animal, and the nature of human dyads in their enormous diversity, it is imperative to understand the characteristics of the channels, or, more generally, of the media of transmission, that connect, singly, in alternation, or in concert, the source of any message with its destination. One must, additionally, take full cognizance of the structure of every intraspecific and interspecific message repertoire and the methods of coding peculiar to each, as well as the crucial role of the bionomic context of every communicative act. Senders and receivers, codes and messages, channels, and the pertinent ecology within which all of these function are known in only the most fragmentary fashion, here a piece of mosaic from the macroregion of the ethologist's field or Hediger's circus, there a piece from the microsphere surrounding the laboratory bench. The numerous and varied bits are, however, hardly ever assembled into a coherently framed ethogram. Our new state-of-the-art handbook, *How Animals Communicate* (Indiana University Press, 1977), like its more primitive forerunner, *Animal Communication* (1968), endeavored to pre-

sent expert sketches of known channels of communication,
as well as to exhibit, in selected groupings of animals,
how these worked *in vivo*. I count it as my good fortune,
and a blessing for us all, that Jack Hailman got carried
away with his assignment to contribute the chapter on
visual communication to the 1977 collection and failed to
follow the publisher's strictures with respect to length.
His excess led to the inspired idea of this additional,
separate monograph, which turns out not only to be the
first book of its kind, but one that is bound to become
an incitement as well as a model for comparable works on
every other known major channel employed in the animal
world, doubtless paving the way for an eventual synthesis
incorporating the long awaited typology of terrestrial
communication at large.

One chapter was excluded by mutual agreement: the
one on optical signs in man. This discussion is of paramount interest, coming from this gifted and unusually
broad-gauged investigator of animal behavior, and will
appear as a long article in *Semiotica*. The mechanisms
involved in human optical communication have very deep
phyletic roots, considering the evidence, for instance,
that animals share with man certain classic optical illusions, such as the Edgar Rubin type of figure-ground
reversal effect. Hailman's work is situated in the
great tradition extending from Hermann von Helmholtz to
R.L. Gregory, but he brings it to an added--and, to my
mind, indispensable--dimension of sound evolutionary perspective. It is amply clear that optical processes in
animals and man constitute a kind of semiosis which has
as its source the regulation, or, to use Cannon's term,
with its curiously old-fashioned ring, homeostasis, of
living organisms in nature and the stability of societal
units both in nature and in culture. Organized systems
(bodies, societies) tend to restore their equilibrium by
the intermediary of certain devices following feedback
stimulation from an unstable environment. One of Hailman's principal tasks is thus to spell out how this particular kind of physiological adaptation frees all of us
who can see from the perennial fluctuations of (as Claude
Bernard might have put it) our *milieu extérieur*. His a-
chievement lies in having delineated the universal aspects
of optical communication without, however, having glossed
over the quite disparate biological phenomena, those subtle
but fundamental disparities that separate species from species, and man from the rest of the animal world.

OPTICAL SIGNALS

Animal Communication and Light

>...the actual physical...structure
>of signals may be determined by a
>number of factors such as the phys-
>ical laws regulating the transmis-
>sion of signal energy, locatability,
>the functional properties of sensory
>organs, the physical characteristics
>of the habitats in which communica-
>tion occurs, and prey-predator re-
>lationships. This fact implies that
>there is some degree of predictabil-
>ity in design features of signaling
>systems. --Konishi (1970)

Chapter 1

INTRODUCTION

1

There are two things which I am confident I can do very well: one is an introduction to any literary work, stating what it is to contain, and how it should be executed in the most perfect manner... * --Samuel Johnson.

The male redwinged blackbird has flaming red epaulets conspicuously displayed to other blackbirds during the reproductive season *(fig 1-1)*. Why is this signal-patch red instead of being green or some other color? Why is it located on the shoulder? Why is it the shape and size that it is? I cannot answer such questions--yet. What I can do in this volume is to collect and organize some strategies for finding the answers.

The analysis of anything, including optical signals used in animal communication, involves breaking the thing into its components for examination. Analysts often fail to make explicit their methods and biases, leaving the reader to guess at them as the analysis proceeds. Such guessing is particularly detrimental to interdisciplinary endeavors such as the analysis of animal communication, which involves physics, ethology, cybernetics, semiotics, sensory physiology, psychophysics and a host of other disciplines in the natural and social sciences. I have therefore dared at the outset to articulate my approaches to scientific epistemology.

The other purpose of an introduction is, as Johnson says, to state what the work is to contain. I suppress

---

*Quote continued at the outset of the final chapter.

the urge to state what it does not contain (therein lies the challenge), and provide instead the expected road-guide.

Fig 1-1. Optical signal of a red-winged blackbird, in which the male prominently displays red epaulets (after Nero, 1956).

Epistemology

Scientists are fond of saying that science is a verb, not a noun. The important thing is to do it, but it is also important to make clear what one thinks he is doing. Scientists do not agree on "the" scientific method, so I present "a" scientific method--namely the one I think I use. I begin with metaphysics, move through traditional concerns of hypotheses and their tests, sketch the roles of mathematics, consider the four biological determinants of behavior and end with discussing the comparative method.

existence and cause

Science begins with philosophy, specifically with
metaphysics. No one has solved to universal satisfaction
the ontological question "why do we believe things exist?"
or the cosmological question "what is cause?" Science
sidesteps both questions, treating them rather than answering them.

We human animals believe things *exist* because we receive certain sensory data that are consistent with belief in some world apart from us. When sensory data dis-
agree--data available to two different persons or two different sets of sensory data to the same person--doubt en-
sues. Thus, one fails to believe in the reality of things
perceived in hallucinations and dreams.

Can one tell, however, whether things exist when
there are no senses (or extensions of senses such as cam-
eras and tape recorders) to record data? Bertrand Russell,
who was fond of the concept of the limit once he had made
it precise, suggested a thought-experiment as follows.
Does a chair exist when there is no one in the room to perceive it? Leave for an hour and return; the chair is still
there. Then leave for half an hour; no change upon return.
Then leave for 15 min, 7.5 min, *etc.* No matter how arbitrarily short an absence is chosen, the chair is always
there upon return. In the limit of indefinitely short ab-
sences, then, the chair remains: it is "always" there and
hence must have a reality apart from our sense data. We
really believe in the existence of things not because of
"proofs" like Russell's, but because to believe leads to
no inconsistencies in our lives and thinking.

Science employs a more satisfactory way to sidestep
the problem of *cause*. If I push on Russell's chair and it
moves, I am tempted to say that my pushing caused the movement. I become really convinced of this explanation, however, when I see the chair fail to move when I do not push,
and when I see the chair always move when I do push. In
short, we infer cause from a constant conjunction of temporally ordered variables, such as first I push, then the
chair moves. One may, of course, investigate the example
in greater detail, measuring lever forces in my arm, fric-
tional drag of the chair's legs on the floor, or even the
molecular and atomic events that transpire. In the end,
however, one merely emerges with a more complete descrip-
tion of a whole chain of temporally ordered variables.

Science owes to the physicist P. W. Bridgman (*e.g.*, 1927) the doctrine that relationships among measurable variables have no reality apart from the operations by which the data are obtained. Science is concerned with such *operational* relationships: observing variables and discovering consistencies in their values relative to one another. Whether or not there is anything more fundamental about causation than constant conjunction of temporally ordered, observable variables is a problem for philosophy rather than science.

epistemological cycles

Any representation of how observable variables are related may be called a *model*. Some special types of models are familiar: physical objects called models are used by marine architects to see how hull design (one set of variables) affects speed through the water, amount of wake created and so forth (another set of variables). Such *physical models* are widely used in experimental analyses of animal communication (see Hailman, 1977a, esp. figures 14 and 18). Another familiar model in science is the *mathematical model*, an abstract statement of relationships among variables expressed in certain arbitrary symbols that are, often literally, Greek to many readers. However, the notion of a model need not be restricted to these extremes: a model is any expression of suggested relationships among observable variables.

One may classify models according to the degree of trust placed in them by the scientific community (Walker, 1963). A *hypothesis* is a well-stated but relatively untested model. A *theory* is a model that has survived sufficient testing to merit widespread interest. Finally, a *law* is a model so well tested that it is unlikely to be overturned completely by new data. Further testing may show the law to be a special, included set of relations within some broader model--as the laws of physical mechanics are now a special case of relativistic models. Hypothesis, theory and law mean other things in other contexts, of course, and one should also admit as models those very tentative statements we call ideas, hunches, suggestions and so on.

A model in science accrues trust by surviving empirical tests, but it is rarely possible to test a model as

a whole. Rather, one employs *deductive reasoning* to derive a *prediction* that *must* be true if the model is true.
It is the prediction from the model that is measured a-
gainst the meterstick of reality: if it does not match
reality, the prediction is false, and so must be the model (assuming the deductive processes were sound). If the
prediction is true (matches reality), however, it cannot
mean the model is *necessarily* true: false models can provide true predictions. For example, the model that the
sun revolves around the earth from east to west correctly
predicts that the sun as seen from the earth rises in the
east and sets in the west. (Other predictions from the
same model do not survive empirical tests, of course.)
Therefore, an empirical test can never prove a model; it
can only fail to disprove it.

The prediction from the model instructs the scientist to collect certain *data*. This she does by *observation*, whether or not a formal "experiment" is performed.
Niko Tinbergen once said in a quote I cannot relocate that
if the observer of animal behavior is patient enough,
Mother Nature will perform the experiments. Various combinations of variables occur spontaneously, so one does
not necessarily have to insure these combinations by ma-
nipulations in the laboratory. Formal experiments do,
however, help isolate the variables of interest so that
one may observe their relationships with greater confidence
that no extraneous variables were influencing the ones of
interest.

One is left with the task of *comparing* the data with
the prediction in order to reach a *decision* as to whether
the two are the same or different. If they are the same,
one goes on believing for the moment that the model correctly predicts observable reality, and ideally devises
some new prediction to be tested. Each time a model correctly predicts reality, one places more confidence in
the model. On the other hand, should the data fail to
match the prediction, one must conclude that the model itself is false. In that case, a new model must be erected
and tested.

The processes for generating models are called *induction*, but these processes are nothing like the deductive logical processes used to derive predictions from the
model. My own belief is that the usual college definitions
of induction and deduction are gobbledygook. Deduction
is not best described as a process of reasoning from generalities to specifics, although that may sometimes be

the case if "generalities" and "specifics" can be satis-
factorily identified. Deduction is, rather, the rearrangement of knowledge contained in the model and its
assumptions. The defining criterion for deduction is
that if the model is true in any sense, the prediction
deduced from it is true in the same sense. Induction,
on the other hand, is poorly described as reasoning from
specifics to generalities. It is doubtful that inductive
processes are best termed reasoning (which connotes some
lawful, logical process), and in any case they do not
necessarily begin with specifics and end with generalities.
It seems best to state that induction is the creative process of science, a potpourri of mental processes whose
end point is a statement about the relationships among observable variables. Induction is as mysterious as the
creation of a great painting or symphony, and like ar-
tists and composers, scientists differ in their creative
abilities.

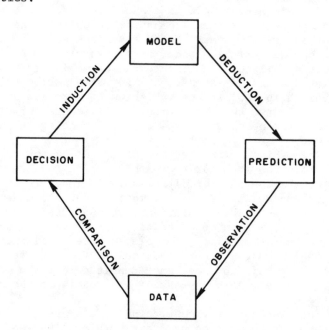

Fig 1-2. *The cycle of scientific epistemology, with processes indicated by arrows and results of processes indicated by boxes to which the arrows point. Mathematics
plays a role in all processes.*

Each small step in this cycle of epistemology could be the subject of a book; I condense the summary into *fig 1-2*. It is my view of how science as a whole evaluates its structure of knowledge, not necessarily how an individual scientist proceeds (Hailman, 1975). Not every individual scientific study will complete the entire cycle. Theoreticians may concern themselves simply with providing models; others may ingeniously manipulate models deductively to provide testable predictions; empiricists may concentrate almost exclusively upon collecting relevant data; and others may devote their lives to careful scrutiny of data to see whether these do or do not match the predictions. In some cases, one scientist will complete the entire cycle, perhaps several times, testing and retesting some model, or continually producing better ones as previous ones fail to predict reality.

This volume on animal communication and light is concerned primarily with creating very tentatively expressed models relating various factors to the characteristics of optical signals. It is a statement of future directions in creating models, deriving predictions from them and testing those predictions.

mathematical roles

Mathematics is not, as often asserted, the language of science—because so many of us have failed to master that tongue. When biology speaks the language with the same fluency as physics, science will be the richer. The emphasis of mathematics in science is, however, often misplaced. Popularizations of science emphasize the precision with which measurements of natural phenomena are made in the observational process of *fig 1-2*. Scientists themselves often pay most attention to statistical comparisons that lead to the decisions as to whether or not data fit the prediction. Nor can the mathematical thinking of Einstein, Bohr or others be denied an important role in their inductive processes. In my view, though, it is in the deductive processes that mathematics plays the most critical role in science.

The recurring problem with grand schemes purporting to explain animal behavior lies with the impossibility of deriving testable predictions from them. Readers suffer through long arguments about the instincts, motivations or approach-withdrawal processes of animal behavior without discovering how these schemes might be put to empirical test. Scrutiny reveals that most such schemes are incapable of falsification, and hence do not deserve the approbation of "model" at all.

In other cases, potentially testable models remain unevaluated because the deductive processes leading to predictions are difficult or faulty. Here, mathematics plays its crucial role in forcing articulation of the model in symbolic form so that rules of mathematical deduction may be employed to derive predictions from it *(fig 1-2)*. Taken broadly to include symbolic logic or its equivalent of Boolean algebra, mathematical deduction makes scientific epistemology possible.

The characteristic of valid deduction is that if the model be true, the prediction *must* be true. If the model be false, valid deduction leads to predictions that may be either true or false (in the sense of matching or failing to match empirically gathered data). It is not difficult to appreciate that *in*valid deductive processes may deliver either true or false predictions, regardless of whether they are derived from true or false models. Therefore, from the viewpoint of the scientist as the decision-point in *fig 1-2*, there exists a logical truth-table that allows him or her to evaluate the truth-value of the model being tested, as shown in *table 1-I*.

Table 1-I

Truth-value of Models, Dependent upon Validity of Deduction and Comparison of Predictions with Data

	PREDICTION	
DEDUCTION	=data (T)	≠data (F)
valid	T or F	F
invalid	T or F	T or F

It is evident from *table 1-I* that if the scientist's deduction is invalid, or if its validity is in doubt, *nothing* can be learned about the truth-status of the model

regardless of the empirical outcome of testing. For this
reason the entire scientific edifice balances on the necessity for valid deduction--validity that must be established by inspection of the deductive steps *per se* without reference to empirical aspects of the scientific problem. Such validity can be established only when the deductive processes may be inspected for strict conformity
with the rules. The best set of rules that exist for un-
erring deduction are those of mathematics.

Finally, one may note from *table 1-I* that if the deductive process is valid, one may with certainty falsify
a model when there is not a match between prediction and
empirical data. When the match exists, the model itself
may still be false: there is no logical way to be sure.
Repeated testing of different predictions, each in turn
failing to reject the model, is the sole method for increasing confidence in it.

behavioral determinants

It is crucial to understand that the question "why
does the animal do that?" is actually several different
questions. There are four classes of models that explain
animal behavior and they are mutually compatible (Tinbergen, 1963; Hailman, 1976a). Because the act of communicating is behavior, there are four kinds of answers to
questions about how communicative behavior is determined.

First, one may ask how internal and external factors
combine to dictate the behavior that we as outsiders observe. This classical concern of motivational and sensory studies may be called the *dynamic control* of behavior. The mere ordering of descriptions of behavior into
coherent schemes constitutes the first model of dynamic
control, later to be refined by sensory and motivational
experiments, and perhaps even physiological studies of
control model of the communicative act.

Second, one may ask how antecedent events in the life
of the animal brought about the control properties of the
behavioral end-point one observes. The two factors that
control such *ontogenetic development* are the genetic endowment from the parents and the environmental conditions
in which the animal develops. The study of ontogeny includes dealing with the classical nature-nurture question
that still arises in modern times as an "instinct-learning"

dichotomy, although an operational approach to the problem quickly alters its formulation (Hailman, 1976b). To
date, rather little is known about genetical and experiential factors in optical communication of animals (Hailman, 1977a), and unfortunately little more can be added
in this volume.

The third fundamental behavioral determinant focuses
upon the population of animals rather than its component
individuals. Many patterns of behavior including communication persist from generation to generation. A high
correlation between the behavior of parents and their offspring may be due to the similarities in their genes or to
the similarities in their rearing environments, including
the cultural environment (ontogeny, above). In most cases,
one assumes that such *preservation* of behavior from generation to generation is maintained by natural selection
or closely related processes such as kinship selection
(see Brown, 1975 and Wilson, 1975).

Because natural selection is a primary cause of behavioral preservation, this determinant is often referred
to as the "adaptive significance" or "selective advantage"
of behavior. I myself have frequently referred to the
"adaptive function" or simply "function" of behavior (*e.g.*,
Hailman, 1967a, 1976a, 1977a). However, the term "function"
is unfortunate: it narrowly connotes only processes of population genetics to the exclusion of experiential factors,
and it has other meanings in biology (Hailman, 1976a).
For example, when one discovers *how* something works (control, above), one often says it "functions" in a certain
way. From the analysis of control one then guesses at the
selective advantages a behavioral pattern confers upon its
practitioner, so that function refers to two related but
distinct behavioral determinants. Finally, some behavioral patterns may persist in populations by mechanisms other
than natural selection, so not every behavioral pattern
has an assignable adaptive function (Hailman, 1977b).

The fourth and last class of behavioral determination
concerns antecedent events in the history of the population,
which may be called the *phylogeny* of behavior. To con-
temporary biologists the word phylogeny is nearly synony-
mous with evolutionary history, but I use the term in its
older and wider sense of simply the history of the population. Therefore, phylogenetic considerations include the
origins of culturally transmitted behavior and hence complete a comprehensive albeit general scheme for investigating behavior.

Table 1-II

Classes of Behavioral Determinants

	immediate cause	antecedent origin
individual organism	CONTROL	ONTOGENY
population of organisms	PRESERVATION	PHYLOGENY

Table 1-II summarizes the scheme of behavioral determinants in pointing out that at both the level of the
individual organism and the population of which it is a
part, there is a class of immediate *causes* of behavior as
well as a class of antecedent *origins*. One cannot, in my
view, claim to understand any behavior, including communicative behavior, without understanding all of its simultaneous and interacting classes of causes and origins.
Unfortunately, when animal communication by light is
scrutinized systematically according to *table 1-II*, the
review of knowledge and understanding turns out to be
short (Hailman, 1977a).

comparative method

The cycle of epistemology *(fig 1-2)* may be applied
directly to all the classes of behavioral determinants
(table 1-II), but it is not always obvious how this ap-
plication should be accomplished. One may experiment directly with the dynamic control of communicative behavior
by manipulating external factors such as social signals
and other stimuli, or by manipulating internal factors
such as hormonal balance (see Hailman, 1977a). Similarly,
the ontogenetic determinants of communicative behavior may
be studied directly by manipulating the genetical endowment or the rearing environment of the developing individual. It is more difficult to study directly the determinants of communicative (or other) behavior at the population level. To do so, biologists frequently employ the
less direct strategy of the comparative method.

The *comparative method* utilizes comparisons among

(sometimes within) species of animals to test predictions
concerning the preservation and phylogeny of behavior. In
its most powerful form four groups of animals are compared
by this method: two evolutionary or cultural lines whose
populations live in each of two environmental situations
(table 1-III). The behavioral characteristics of the

Table 1-III

Groups for the Comparative Method

evolutionary or cultural line	environmental situation	
	a	*b*
line *a*	group (*a*,*a*)	group (*a*,*b*)
line *b*	group (*b*,*a*)	group (*b*,*b*)

groups of populations are then inspected for similarities
and differences. If traits are similar in groups (*a*,*a*)
and (*b*,*a*) of the table, but different from traits in groups
(*a*,*b*) and (*b*,*b*) which are similar to one another, then the
differences correlate with differences in the environmental
situation. One attributes the correlation to mechanisms
of preservation that are tied to the immediate environmental situation. Alternatively, if groups (*a*,*a*) and (*a*,*b*)
show similarities, and groups (*b*,*a*) and (*b*,*b*) show similarities of a different kind, then the traits under investigation may be attributed to phylogeny: historical determination due to common evolutionary or cultural descent.
    Although evolutionary biologists think almost uncon-
sciously in terms of comparative studies, persons differently trained have difficulty applying and understanding
the method. The difficulty is partly due to the idealized
scheme of *table 1-III*: rarely can all four cells of the
table be filled. For example, E. Cullen (1957) suggested
that the black neck-mark on the chick of the kittiwake
gull is an optical signal that inhibits parental aggression.
This cliff-nesting species is forced into parental-off-
spring propinquity for a longer period than in surface-
nesting gulls, where chicks are soon able to roam over the

relatively large reproductive territory of their parents.
Cullen's suggestion is a model in the sense of *fig 1-2*:
it predicts that forced propinquity of cliff-nesting favors selection for a particular kind of signal. Let the
kittiwake represent group (*a,a*) in *table 1-III*. Then
closely related gull species that do not nest on cliffs,
represented by group (*a,b*), should lack the neck-band signal; as Cullen points out, this is true for all known sur-
face-nesting gulls. Armed with her model, I investigated
the cliff-nesting swallow-tailed gull of the Galapagos
Islands and found the older young possessed a black neck-
band as predicted (Hailman, 1965), providing another example of group (*a,a*). The consistent difference between
groups (*a,a*) and (*a,b*) fails to reject Cullen's model of
preservation: one continues to believe that the aggression-
inhibiting signal is selected for by forced propinquity
of parent and offspring.

Alternatively, other optical signals of these same
gulls have been subjected to considerable ethological
scrutiny (N. Tinbergen, 1959). When their signals are
compared with those of other groups of birds (line *b* in
table 1-III), consistent differences between gulls and
other birds are evident, regardless of the environmental
situations in which the species live (columns of *table 1-
III*). Tinbergen rightly concludes that at least certain
characteristics of optical displays are due primarily to
common evolutionary ancestry.

Execution

It is one thing to think about epistemology, a different thing to execute it. This volume must be content
primarily with assembling disparate factors that may contribute to models about animal communication and light.
In some cases, it is possible to articulate specific models about optical signals, and in a very few cases to mar-
shal sufficient anecdotal evidence for a preliminary test.
The principal concern is to understand why optical signals
have the characteristics they do. To appreciate how this
concern is maintained, the plan and mechanics of the volume are provided here.

strategic gameplan

The overall strategy is straightforward and charted
in such a way that its outline could be followed in scru-
tinizing chemical, acoustical or other signals having
nothing to do with light. First, a framework of the communication process itself is provided as a synthesis combining some of the notions of ethology, cybernetics and
semiotics (ch 2). Then in successive chapters the characteristics of the communication channel (ch 3), the sender (ch 4) and the receiver (ch 5) are scrutinized for
factors that constrain the design of optical signals. Next
the communication of misinformation, in the traditional
sense of concealment and mimicry, is reinvestigated from
the viewpoint of optical principles employed in visual deception (ch 6). The design of optical signals to minimize
the effects of environmental noise then places communication in the ecological context (ch 7). The analysis cul-
minates with considerations of how the qualitative kind
of information sent helps to determine the design of optical signals (ch 8). The volume concludes (ch 9) with a
few comments on the problems and prospects of studying optical signals.

tactical rules

The book is structured linearly and best read straight
through, although various aids are used to facilitate ref-
erencing back to earlier chapters. Each chapter follows
a similar plan, the plan and mechanics being obvious upon
reading. The problems of mathematical notation prove
crushing. Because one quickly exhausts the English and
Greek alphabets, two compromises were made to avoid eso-
teric type-symbols: many relationships are stated in words
rather than equations and symbols remaining may be used
differently in different chapters, although they are used
to mean only one thing within a chapter. Equations are
listed in the contents for ready reference and the symbols
themselves appear in the index. The terminal bibliography
has been used as an index to citations of references in
the text for further aid.

Some mathematical operators may be unfamiliar. I follow Batschelet's (1975) sensible suggestion of using the
proportionality operator ($\propto$) when constants and parameters

of an equation are irrelevant to the point being made.
Thus, $y = ax+b$ may be rendered simply $y \propto x$ when the values
of a and b are not of central concern.   There is, alas, no
widely used symbol for the monotone operator, so it is not
without a touch of irony that one is invented for a volume
on optical signals. When the value of y never decreases
with an increase in the value of x, this function may be
written $y \uparrow x$, unless the increase is proportional and may
be expressed more precisely. Finally, when the value of
y depends upon the value of x, but not necessarily in a
monotonic relationship, the function is denoted $f: x \mapsto y$
(Batschelet, 1975), and one states that "x maps to y."
This means precisely the same thing as the older notation
$y = f(x)$. The mapping or functional operator ($\mapsto$) was devised to prevent confusion with logical implication ($\Rightarrow$)
and convergence ($\rightarrow$), all three being denoted formerly by
the last symbol. Higher mathematics are not employed, although the reader requires some familiarity with basic
notions of set theory, probability, algebra, geometry,
trigonometry and logarithms.

Overview

My view of attacking scientific problems is basically
that of a "logical empiricist," striving to formulate re-
lationahips among observable variables that may be tested
against reality. A real understanding of communicative
behavior necessitates study of control, ontogeny, preservation and phylogeny, integrating these determinants in a
coherent framework. The primary concern of this volume is
to predict characteristics of optical signals; the comparative method is particularly useful in checking tentative
models of the sort to be proposed. The essay proceeds
linearly, with considerations of the communicative process,
the optical channel, the sender and the receiver; then this
background is applied to analysis of optical principles in
deception, in combatting environmental noise and in transferring different kinds of information. The aim is to provide a gameplan and some rules; the reseach game itself is
a continuing endeavor.

Recommended Reading and Reference

I use these sections to suggest some general works
that may provide a more complete explication of the chapter's subject matter, a different viewpoint from my own,
or just stimulating reading. The choices are, of course,
highly subjective. In addition, the *Appendix* lists articles from *Scientific American* in which one may obtain
authoritative but non-technical reports of relevant scientific findings and viewpoints.

Alfred North Whitehead and Bertrand Russell have obviously influenced my views, but their joint effort, *Prin-
cipia Mathematica*, is no fodder for beginners. An interesting start might be Russell's (1945) *History of Western
Philosophy*, a delightful if personal view of our intellec-
tual heritage. The name of Percy W. Bridgman is conspicuously absent from Russell's writings, and I recommend
restoration of a balance with *The Intelligent Individual
and Society* (Bridgman, 1938) or a more technical work (*e.g.*,
Bridgman, 1927). Karl Popper's (1959) *Logic of Scientific
Discovery* is useful, and one might delve into writings of
Einstein, Eddington, Wittgenstein, Kuhn and others, as well
as classical philosophers. I find Plato the most imagin-
ative of all, and if my conception of a scientific model
seems like his "shadow of an image" projected on the wall
of a dark cave, so be it.

Chapter 2

COMMUNICATION

No pleasure is fully delightsome without
communication. --Montaigne.

Communication is the characteristic of social behavior. Being social animals ourselves, we human beings have
a practical stake in understanding communication, and in
any case, in order to appreciate the design of optical
signals we must have some notion about the communicative
process. Three principal disciplines have contributed to
the study of animal communication: ethology, semiotics and
mathematical information theory, a branch of cybernetics.
Ethology, the science of animal behavior, has concerned
itself primarily with the evolution and adaptiveness of
social signals *(ch 1)*, although its general framework includes the dynamic control and ontogenetic development of
communicative behavior. *Semiotics,* the study of signs,
has concerned itself primarily with the formal structure
of the communicative act. *Cybernetics*, particularly information theory, has concerned itself primarily with
quantitative measures of communication. In this chapter
I attempt to combine these three disciplines in a framework useful for studying the design of communication signals that utilize light.

Types of Communication

Communication may occur when a *sender* initiates in
some *channel* a physical disturbance (the *signal*) that is
detected by a *receiver (fig 2-1)*. For example, a female

-21-

Fig 2-1. A simple communicational system, in which the
communicants are represented as boxes connected by a channel represented as a line. Communication occurs when information is transferred via a signal in the channel from
one communicant (the sender) to the other (receiver) as
indicated by the arrowhead.

firefly (sender) emits a pulse of light (signal in the
optical channel) that is seen by a male firefly (receiver).
Actually, mere detection by the receiver is not sufficient
for communication, and later in this chapter a comprehensive and precise definition of communication will evolve.
One might classify types of communication according to the
communicants (sender and receiver), according to the channel, or according to other characteristics of the process.
This volume concerns a limited group of communicational
types, delimited as follows.

communicants

Senders and receivers may be animals, plants or even
non-living entities such as computers. A classification
based on the kind of communicants may be represented by
a Venn diagram (fig 2-2) in which the universal set is
that of all communicants. The major subsets are living
senders and receivers, and their included subsets of
animals. Each logical space in the diagram represents a
kind of communication system, and the spaces in fig 2-2
are numbered for convenience. Logical space (1) is abio-
logical communication, in which neither communicant is
alive, as in signaling between two computers. The remaining spaces are types of biological communication, in which
at least one of the communicants is alive. Included within biological communication are spaces (5) through (9),
zoological communication, in which at least one of the
communicants is an animal. The focus of this volume is
logical space (9), in which both communicants are animals:
animal communication.

COMMUNICATION

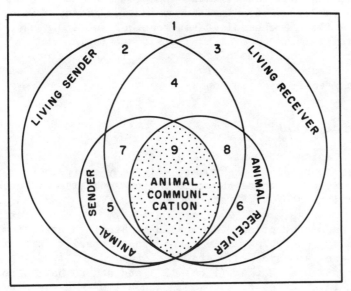

Fig 2-2. Types of communicants may be used to classify communication. Each logical space in the diagram represents a different type of communication, numbered for reference in the text. Space (9) is the focus of this volume.

That animals may communicate with nonliving communicants and with plants is readily demonstrated by example. A star (sender) may send light (signal) detected by a nocturnally migrating songbird (receiver), which utilizes the signal to adjust its direction of flight (logical space 6). Both Smith (1968: 45) and Marler (1968: 103) discuss communication between plants and animals (spaces 7 and 8), as when the flower's nectar guides direct a bee to nectar (space 8).

channels

An obvious basis for classifying communication is by the physical channels through which signals are sent. To understand animal communication it is necessary to study

all the channels simultaneously: optical, acoustical, chemical, tactual, *etc.* There appears to be no thorough and
general study of animal communication in any channel, much
less all the channels taken together, so it is useful to
partition the general analysis of communication by channel.
Although the focus of this volume is the optical channel,
the strategy of analysis may provide a model for similar
analyses in other channels.

animal communication

Animal communication as defined by *fig 2-2* includes
a broad range of phenomena, which may be further subdivided
according to three characteristics emphasized in ethological literature. First, many authors (*e.g., Wilson, 1968:
98; Tavolga, 1968: 273) draw attention to communication between different species, as in symbiosis, parasitism, predation, *etc.* This volume concerns primarily communicants
of the same species: *conspecific communication.*
Second, Marler (1968: 103) and others have stressed
the importance of "two-way" exchange of signals between
the communicants, a notion I call *communicative reciprocity.*
Not all conspecific communication is reciprocal in this
sense; for example, one animal of a group may give an a-
larm call to which others respond only by fleeing. T. Johnston *(pers. comm.)* points out that linguists apply a notion
of reciprocity to signals: if two individuals utilize the
same set of signals or vocabulary of elemental units, these
units possess a reciprocal quality. That reciprocity is a
special case of communicative reciprocity in general, a
case that may apply to certain kinds of animal signaling
(*e.g.,* aggressive threat signals of two individuals) but
not to others (*e.g.,* differences in signals used by male
and female in courtship communication).
Finally, many ethologists emphasize the role of natural selection in structuring signals to maximize their
efficiency and reliability in transferring information
(*e.g.,* Klopfer and Hatch, 1968: 31-32; Smith, 1968: 44-45;
Marler, 1968: 103). Julian Huxley (1923) referred to highly evolved signals as "ritualized," a term ethologists apply to any behavioral patterns whose characteristics have
been evolutionarily influenced because the patterns serve
as signals. Although difficult to identify in practice,
such *ritualized communication* has been a major focus of
ethology.

ANIMAL COMMUNICATION

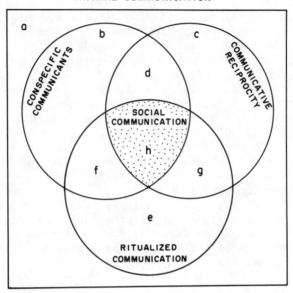

Fig 2-3. Three properties of communication (circles) define various types of animal communication (see space 9 in fig 2-2). The logical spaces are denoted by letters for reference in the text, and the focus of this volume is space (h).

Figure 2-3 shows these characteristics of conspecificity, reciprocity and ritualization in a Venn diagram that defines eight logical spaces, all of which are subsets of space *(9)* in *fig 2-2*. Only in certain cases will it be easy to decide whether or not some instance of communication belongs unequivocally to a particular set, so these should be regarded as "fuzzy sets" presented only for the purpose of pointing out the principal kind of communication treated in this book. Space *(a)* might include the example of a hidden songbird watching a hawk soar overhead, *(b)* such a songbird watching one of its conspecifics moving about the environment, *(c)* the hawk chasing the songbird, and *(d)* a group of such songbirds maneuvering in a coordinated flock during flight. The remaining spaces involve more highly ritualized communication, and an example of *(e)* might be the display of eye-spots on the

wings of a moth confronted with a predatory songbird.
Space *(f)* includes the songbird giving ritualized threat
display to a conspecific that shows no response, and space
(g) includes the same bird threatening a member of another
species in a mixed-species foraging flock, when the other
bird responds in some way to the threat. Space *(h)*, the
mutual intersection of the three defining sets, I call
social communication: reciprocal, ritualized communication
between conspecifics. Social communication is the primary focus of this volume, and is roughly equivalent to
Marler's (1968: 103) "true communication" and Smith's (1968:
45) "social signal function."

Only certain groups of animals combine sufficient
complexity of social behavior with keen enough vision to
render their communication in the optical channel interesting and challenging to ethologists. This volume reflects the literature on those groups: insects, some other
arthropods (such as spiders and crabs), cephalopod molluscs
and most classes of vertebrates (bony fishes, amphibians,
reptiles, mammals and birds).

Signals and Information

Because communication is the transfer of information
via signals, it is necessary first to make precise the notions of signal and information. *Semiotics*--or in its o-
riginal parallel with "logic," semiotic--is the study of
signs, which is an axiomatic framework and vocabulary for
analysis rather than a predictive theory of communication
(see Johnston, 1976 and Sebeok, 1968). Cybernetics, particularly the branch called the mathematical theory of communication or *information theory*, deals quantitatively with
measuring the amount of information transferred in communication. The objective of this section is to extract certain notions from both disciplines and combine them into
a basic framework useful for analyzing the characteristics
of optical signals.

signals

Charles Sanders Peirce, founder of semiotics, con-
ceived of communication as taking place whenever a *sign*
stands for something (its *object* or designatum) to some-

body (the *interpreter*) in some manner (the *interpretant*
or interpretation of the sign). C.W. Morris (*e.g.*, 1946)
translated these general notions into the physical world,
in which a *sign-vehicle* stands for a *referent* and produces
a *response* in the interpreter. Marler (1961), W.J. Smith
(*e.g.*, 1968) and other ethologists have used Morris's
framework for analyzing animal communication, and recently
there have been attempts to clarify and extend such uses
(*e.g.*, Stephenson, 1973; Johnston, 1976). In a simple
example, the black mustache mark of the adult male common
flicker in eastern North America (see figure 10 in Hailman, 1977a) stands for "adult maleness" (the referent) to
other flickers (the interpreters) and produces observable
effects on those interpreters. When Noble (1936) painted
a mustache mark on a female flicker, her mate aggressively
drove her away until Noble recaptured her and removed the
mark.

Semiotics distinguishes between the *signal*, which is
a collection of physical disturbances, and the included
sign-vehicle, which is that aspect of the signal producing
the response. For example, the Carolina anole is a small
lizard in which the male extends his red throat-skin,
called the dewlap, and makes certain display movements.
Females watching a displaying male through glass partitions
undergo ovarian development: a response to the optical signal. Crews (1975) found that surgical prevention of the
movement abolished the response whereas changing the dewlap's color to black did not affect the response. The
color is therefore not part of the sign-vehicle that affects
female ovarian response. Ordinarily, ethologists do not
know what part of an optical signal is producing the response; experimentation by altering the signal or by using
models is required (see citations in Hailman, 1977a). It
is also possible that different aspects of the same signal
are different sign-vehicles, but this problem has been lit-
tle-studied in optical signals of animals.

In this book, I refer to the male flicker's mustache,
the male anole's dewlap and other markings and structures--
as well as certain movements, postures and orientations--
as optical signals, even though such usage entails at least
three problems. First, it blurs the useful distinction between the signal and its included sign-vehicle, but since
the distinction cannot be made except in a few well-studied
examples, there is little choice but to gloss over it with
a call for more experimental studies. Second, the mus-

tache mark or dewlap are not transmitted from sender to
receiver: the light reflected from them is transmitted.
Therefore, the markings, structures, movements and so on
might usefully be called "signal-objects," but in general
I shall make the distinction explicit only when confusion
might occur. Finally, it is not always clear how one signal can be separated objectively from others in space or
time, and there are cases in which one cannot specify "a"
signal. Usually, this problem causes little trouble and
may be made explicit when it does.

uncertainty

Information is the reduction of uncertainty, so one
must first make precise the measurement of uncertainty.
Players of the parlor game "twenty questions" quickly
learn the optimum strategy of eliminating half the possibilities remaining with each question asked, which must
be answered yes or no. If an array about which one is uncertain consists of the four members A, B, C and D, the
first question might be: is it either A or B? If the answer be affirmative, one may ask if it is A; if negative,
whether it is C. In either case, the correct item will be
identified in two questions. An array of two items requires but a single question; an array of eight, three
questions; an array of 16, four; and in general an array
of n items requires H questions when

$$2^H = n ,$$

which may be rearranged in logarithmic form as

$$H = \log_2 n .$$

$$(2.1)$$

The quantity H is the *uncertainty* (or *entropy*) of the array of n equiprobable items, and the units are called *bits*
(*bi*nary dig*its*).

It is easy to show that the binary strategy of guessing is the most efficient way to play the parlor game. For
example, if the first question were "is it A?" the guess
would be correct only in one-quarter of the games. If negative, one might then ask "is it B?" which guess would also
be correct in only one-quarter of the games. The third
question "is it C?" will always remove all doubt in the re-

maining half of the games. The average number of questions required when using this serial strategy is therefore $1(1/4) + 2(1/4) + 3(1/2) = 2.25$ questions/game, whereas the "halving" strategy explained at the outset always
requires exactly two questions. The efficiency of the binary technique helps one to feel intuitively that *eq (2.1)*
is a useful expression for measuring uncertainty in arrays
of equiprobable items.

    Equation (2.1) gives a sensible answer only when the
array is defined. If the task is to identify a particular
playing card drawn from an ordinary deck without jokers,
$H = \log_2 52 = 5.7$ bits/draw. If the task is to identify
only the suit of the card drawn, the uncertainty is less
because there are only four equally probable suits: $H =
\log_2 4 = 2$ bits/draw. If someone removes certain cards
from the deck before making the draw, the array of alternatives cannot be specified unless one knows what cards
were removed; it may be impossible to calculate uncertainty. In most cases of animal communication, one cannot define the array and therefore cannot calculate uncertainty.

    Suppose the items of the array are *not* equally prob-
able, as might happen when someone removes cards from a
deck. Even if one knows which cards remain, *eq (2.1)* can
be employed only if the distinguishable items of interest
(say suit) occur with equal frequency. In order to gen-
eralize *eq (2.1)*, consider the fact that each equiprobable
item contributes exactly $1/n$ portion of the total entropy,
so that the equation may be rewritten

$$H = \sum^{n} 1/n \,(\log_2 n) .$$

Since the probability of the ith item in the array is $p_i =
1/n$, it is also true that $n = 1/p_i$ for the n equiprobable
items. Substituting,

$$H = \sum_i^{n} p_i \,(\log_2 1/p_i) ,$$

from which the fraction may be removed by noting that $\log
1/x = -\log x$, so

$$H = \sum_i^{n} p_i \,(-\log_2 p_i) , \qquad\qquad (2.2)$$

which is Shannon's general equation for entropy (Shannon
and Weaver, 1949), often written with the negation sign
removed outside the summation operator. One must accept

intuitively that *eq (2.2)* delivers useful answers when the
component p_i's are not all $1/n$; the only requirement is
that $\Sigma \ p_i = 1$.

In order to measure uncertainty with *eq (2.2)* it is
still necessary that the array be defined. A canasta deck,
for example, has two ordinary decks plus four indistin-
guishable jokers. The uncertainty facing one attempting
to guess a card drawn at random is: $H = 4/108 \ (-\log_2 4/108)$
$+ \ 2/108 \ (-\log_2 2/108) \ 52 = 0.176 + 5.538 = 5.714$ bits/draw.
If the jokers of one deck were distinguishable from those
of the other, the uncertainty would be: $H = \log_2 54 = 5.755$
bits/draw. The comparison illustrates two points about
uncertainty. First, uncertainty has a maximum value when
the items of the array are equally probable (5.755 > 5.714),
and second, uncertainty is larger with larger arrays (5.755
> 5.700 of a 52-card deck without jokers). When arrays are
of equally probable items, *eqs (2.1)* and *(2.2)* deliver the
same result, and when arrays contain but a single item (n
= 1), uncertainty is zero (no uncertainty if there is no
choice).

information

Information is the decrease in uncertainty as the result of communication, and hence is calculated as the difference between initial (H_0) and subsequent (H_1) entropy:

$$I = H_0 - H_1 , \qquad\qquad\qquad (2.3)$$

with units in bits. Frequently, communication abolishes
all uncertainty, so that $H_1 = 0$, and therefore $I = H_0$.
This is the sense in which authors may use uncertainty
(entropy) and information interchangeably.

The information associated with communication may be
difficult to calculate. Consider surface-feeding ducks
of the genus *Anas*, which have species-specific wing-markings
called specula (see figure 20 in Hailman, 1977a). If there
are eight equally abundant species on a lake, observing the
speculum of a particular individual might be said to communicate $I = \log_2 8 - 0 = 3$ bits/bird. Although this is
the information about species transferred to a bird-watcher,
it is likely to over-estimate the information transferred
to another bird. Each duck probably distinguishes its
species from all others combined, so that the initial un-

certainty by *eq (2.2)* is only: $H_0 = I = 1/8 \; (-\log_2 1/8) +$
$7/8 \; (-\log_2 7/8) = 0.54$ bit/bird. In many cases the ethologist has almost no basis for calculating uncertainties
at all, and hence cannot estimate the information transferred by a signal.

Haldane and Spurway (1954) cleverly showed that it is
sometimes possible to calculate information without being
able to calculate a meaningful value for either initial
or subsequent uncertainties. The honey bee worker returning to the hive from a foraging site may move in a figure
"8" on the vertical comb. The angle of this "dance" with
respect to the vertical correlates remarkably well with
the direction of the food-source with respect to the sun
(von Frisch, 1955). Thus, if the bee moves leftward
through the portion of the dance between the two loops of
the "8" the source is 90° to the left of the sun; if she
moves downward through that segment (the "8" being on its
side: ∞), the source is directly away from the sun. Another worker before sensing the dance is initially faced
with a full circle of alternative directions in which to
search for the food-source, but how many alternative directions are there? If H_0 be measured by the cardinal
directions, the uncertainty is $\log_2 4 = 2$ bits/foray, but
if every degree is recognized as an alternative, the uncertainty is $\log_2 360 = 8.5$ bits/foray.

Haldane and Spurway solved the problem by recognizing
that the difference between the logarithms of two numbers
may be expressed as the logarithm of the ratio of the
numbers: $\log x - \log y = \log (x/y)$. They obtained estimates of the angles at which outgoing worker bees dispersed
after sensing the dance, from these data computed the
standard deviation *SD* (which they call the standard error),
and then employed an equation of Shannon's to obtain an
estimate of H_1:

$$H = \log_2 (SD \sqrt{2\pi\varepsilon}) , \qquad\qquad (2.4)$$

where *SD* is the standard deviation of a normally distrib-
uted population of data and ε is the mathematical constant
2.718..., the base of natural logarithms (Shannon and Weaver, 1949). Therefore,

$$I = \log_2 (\text{circle}) - \log_2 (SD \sqrt{2\pi\varepsilon})$$

$$= \log_2 (\text{circle}) - \log_2 SD - \log_2 (\sqrt{2\pi\varepsilon})$$

$$I = \log_2 (\text{circle}/SD) - 2.05 \ .$$

However the standard deviation is expressed (in degrees, grads, radians, *etc.*), so long as the full circle is expressed in the same units the ratio (circle/SD) is constant and the information transferred has the same value. Gould (1975) points out that the answer obtained by Haldane and Spurway probably underestimates the real amount of information transferred due to technical considerations, but that story is an aside. The example serves to illustrate the usefulness of informational calculations as the "common currency" for comparison among physically dissimilar communication systems. Wilson (1962) later used the same method of calculation to compare dancing of honey bees with chemical trail-following in fire ants.

Usually, the concern is with the *average* amount of information transferred, but sometimes it is useful to calculate the information transferred by a single type of signal when not all alternative signals transfer the same amount of information. Consider the case in which all uncertainty is abolished by receipt of a signal ($I = H_0$). The average information transferred *(eq 2.2)* may be partitioned into its components, each ith signal transferring exactly $-\log_2 p_i$ bits, which is weighted in the average according to its probability of occurrence: p_i $(-\log_2 p_i)$. This means that very rare signals (small p_i's) transfer relatively more information at each occurrence than do common ones, since $-\log_2 p_i$ is inversely proportional to p_i.

In this sense, information is akin to *news* as defined in the *New York Sun* in 1882 by editor Charles A. Dana (or, more likely, his city editor, John B. Bogart): "When a dog bites a man that is not news, but when a man bites a dog that is news." Suppose there are 999 cases a year of dogs biting men and one instance of some canophobe chomping on man's best friend. The average information transferred by reports would have bored Dana and Bogart: $I = H_0 - H_1 = 0.999$ $(-\log_2 0.999) + 0.001$ $(-\log_2 0.001) - 0 = 0.01$ bit/report. The specific report that a man bit a dog, however, transfers 9.67 bits/report. The rumor that the unit of information was named from such animal behavior is, however, without foundation.

It may be confusing logically to use the concept of information in this latter sense applied to individual signals. (Many mathematicians similarly object to the notion of probability applied to individual events.) Attneave

(1959: 6) recommends the use of "surprisal" for entropic
calculations of individual signals, and in the remainder
of this book information implies entropic quantities of
signal-sets rather than individual signals.

Behavior

The third element of the analysis of animal communication is *ethology*, the study of animal behavior. The act
of sending an optical signal is an act of behavior of the
sender, and the only means for judging the transfer of information is the behavior of the receiver. (One may ask a
human receiver if he or she is more certain about something
after receipt of a signal, but the reply is merely an aspect of his or her observable behavior.) Just as cybernetics helps formulate semiotic problems through information theory, it helps formulate behavioral problems through
control theory. This section combines basic notions of
control theory and ethology to create a framework for the
analysis of behavioral control.

behavioral outputs and signals

There is no agreed-upon, fundamental unit of behavior,
although in many species the behavior appears to be composed of a sequence of natural units called *action patterns* (see G.W. Barlow, 1977). For a given analytical
purpose, behavior may be divided into n distinguishable,
mutually exclusive *behavioral outputs*, denoted b_i (where
$i = 1, 2, 3, ..., n$). When the number of outputs is specified, the maximum *output uncertainty* (H') may be calculated from *eq (2.1)*. Ethologists usually recognize fewer
than 50 alternative outputs, and a reasonable upper bound
can probably be set at 1000, so that H'_{max} = 9.97 bits/
output. If the animal can change output every second, its
maximum entropic rate of output could be on the order of
10 bits/s. Information theorists use the same symbols
(*e.g.*, H) to denote uncertainty (entropy) and the entropic
rate; the two quantities must be distinguished by context
or units (bits/occurrence *vs* bits/time).

Certain behavioral outputs, particularly certain kinds
of movements, postures and orientations of animals, serve
as optical signals (see Hailman, 1977a for a review).

Moynihan (1970) surveyed the literature on such display
outputs and found that the average reported number was a-
bout 20 for fishes, birds and mammals; the rhesus macaque
had the greatest number (37), which comes as no surprise
as it is the most widely studied primate. Wilson (1972)
reports that similar surveys of social hymenopterans show
that bees and ants have between 10 and 20 communicative
outputs. If, say, 50 is a real upper bound, then the maximum *entropy of output signals* ($H*$) is on the order of
5.6 bits/output-signal. Of course, there are evident problems with defining the signal-repertoire of any animal,
and this question is taken up again in *ch 8*.

spontaneous transition

From the beginnings of the discipline, ethologists
have emphasized the occurrence of spontaneity in behavior
(*e.g.*, Lorenz, 1950; N. Tinbergen, 1951). Even when the
external conditions are held as constant as possible, the
animal's behavior changes with time: animals behave without "responding" to anything. For this reason, ethologists
rarely refer to the units of behavior as "responses."
The spontaneity of behavior suggests that the performance of an output alters the animal itself in some way,
and that internal change of state in turn affects the future outputs of the animal. Control theory provides a
framework for treating such a problem. Suppose the animal
were a "determinant machine" (Ashby, 1961), such that it
always cycled through a fixed sequence of its n outputs.
One may tabulate the fixed sequence as a matrix of preceding and next outputs, placing a *1* in cells where the
transition occurs and a *0* in cells where it does not *(table 2-I)*. Each column contains exactly one *1* because the
control of the machine is determinant. The particular
machine illustrated by *table 2-I* cycles through the outputs 1, 2, 4, 3, 1, 2, ... indefinitely; its next output
is always uniquely determined by its previous output. Such
a machine shows *Markovian behavior*.
Animals are not simple determinant machines; rather,
they are probablistic machines whose outputs are rarely
uniquely determined. In order to reflect this fact, the
entries of the transitional matrix *(table 2-I)* may be re-
placed with conditional probabilities. Consider the previous output of the animal to be the measure of its inter-

Table 2-I

*Example of a transition matrix
of a simple determinant machine.*

		previous output (b_i)			
		1	2	3	4
next output (b_i)	1	0	0	1	0
	2	1	0	0	0
	3	0	0	0	1
	4	0	1	0	0

nal state of readiness, or *behavioral state,* denoted β_j
$(j = 1, 2, \ldots, \nu)$. For the moment, $j = i$ and $\nu = n$, al-
though that will change when the notion of β is broadened.
The cells of the new matrix *(table 2-II)* are the condition-
al probabilities $p(b_i|\beta_j)$, which may be written more sim-
ply $p(i|j)$. As in *table 2-I* above, the columns of *table 2-
II* sum to unity (the animal must always do something).

Table 2-II

State matrix of a probabilistic machine.

		behavioral state (β_j)					
		1	2	...	j	...	ν
output (b_i)	1	$p(1\|1)$	$p(1\|2)$	...	$p(1\|j)$	...	$p(1\|\nu)$
	2	$p(2\|1)$	$p(2\|2)$	...	$p(2\|j)$	...	$p(2\|\nu)$
	...	...	...	...	...	...	...
	i	$p(i\|1)$	$p(i\|2)$	...	$p(i\|j)$	...	$p(i\|\nu)$
	...	...	...	...	...	...	...
	n	$p(n\|1)$	$p(n\|2)$	...	$p(n\|j)$	...	$p(n\|\nu)$

The notion of the behavioral state (β) may now be
partially generalized to reflect the diversity of ways in
which ethologists operationally judge the animal's state
of readiness. One way to achieve higher predictability of
the next output might be to consider the behavioral state
to be not simply the previous behavioral output, but rather
the sequentially ordered pair of preceding outputs. In
this case, there are $\nu = n^2$ columns of the matrix of *table 2-II*: b_1b_1, b_1b_2, ..., b_1b_i, ..., b_1b_n, b_2b_1, b_2b_2,
..., b_ib_1, ..., b_nb_n. It is still true that the conditional probabilities in each column must sum to unity, since
after each given pair of preceding outputs some next output must always occur. Such a technique may be carried
yet another step, by considering the ordered triplet of
preceding behavioral outputs to constitute the behavioral
state, in which case $\nu = n^3$. This technique of probabil-
istically predicting behavioral outputs from a chain of
previous outputs is referred to as *sequential analysis* or
analysis of *Markov chains*.

semi-Markovian behavior

It is possible to express the output entropy of any
column in *table 2-II* using a modification of *eq (2.2)* for
conditional probabilities:

$$H'_j(i|j) = -\sum_i p(i|j) \log_2 (i|j) , \qquad (2.5)$$

and to take a weighted average of the column entropies as

$$\hat{H}'(i|j) = \sum_j p(j) H'_j(i|j) .$$

By substituting the right side of *eq (2.5)* for $H'_j(i|j)$,
the weighted average can be written parsimoniously as

$$\hat{H}'(i|j) = -\sum_j\sum_i p(j) p(i|j) \log_2 p(i|j) ,$$

which may be simplified by noting that $p(j) p(i|j) = p(ji)$:

$$\hat{H}'(i|j) = -\sum_j\sum_i p(ji) \log_2 p(i|j) . \qquad (2.6)$$

Equation (2.6) is general for *table 2-II*, regardless of
what is taken to be the behavioral state of the animal.
It is now possible to consider quantitatively the ef-

fects of taking longer chains of previous outputs upon predictability of the next output. If the previous or-dered pair of outputs is taken to represent the behavioral state, there are $nv = n^3$ cells in the matrix. Let g be the exponent of n so that there are always n^g cells in the matrix. When only the immediately preceding behavioral pattern (output) is taken as the behavioral state, $g = 2$; when the behavioral state is unknown, or there is only one possible state, $g = 1$; and when the probabilistic distribution of the n outputs is not specified, let $g = 0$.

One now indexes the output entropy (H') by the subscript g, so that H'_{max} calculated before becomes H'_0; the entropy considering only the overall probabilistic distribution of outputs, regardless of behavioral state, becomes H'_1; the average entropy when the previous output is known becomes $\hat{H}'_2$; the average entropy when the preceding ordered pair of outputs is known becomes $\hat{H}'_3$; and so on. An analysis of grooming behavior of house flies by Sustare and Burtt (*in prep.*) yielded these values based on 14 recognized outputs: $H'_0 = \log_2 14 = 3.81$ bits/output, $H'_1 = 3.03$, $\hat{H}'_2 = 1.53$ and $\hat{H}'_3 = 1.33$. This result is typical of many ethological studies, in that the largest drop in uncertainty occurs between H'_1 and $\hat{H}'_2$ (also see Hailman and Sustare, 1973). Whenever $H'_0 \geq H'_1 \gg \hat{H}'_2 \geq \hat{H}'_3$, the behavior is *semi-Markovian*, meaning that the primary observational basis for judging the behavioral state of an animal is its immediately preceding output. The ethological literature suggests that much of animal behavior is semi-Markovian in nature, although communicative behavior may well be an exception.

behavioral states

Two major difficulties with the Markovian approach to behavioral states are the recording and sample-size problems. The approach virtually requires automatic recording methods, which may be difficult to devise and cumbersome to use in the field. If the sequential analysis is to be carried beyond $g = 2$, tremendous sample sizes are required to fill the cells of the transitional matrix. The second problem is alleviated somewhat when behavior is semi-Markovian, so that there are only n^2 cells to the matrix, but if the animal has just 10 recognized outputs, there are still 100 cells to be filled with probability estimates, so even in simple cases the problem is not trivial.

Because of these problems, ethologists use other methods for identifying the behavioral state of an animal. One method is to note morphological indicators of internal states, such as engorgement and coloration of sex skin in baboons (see figure 30 in Hailman, 1977a). Another approach to long-term behavioral states in general is the overall complex of outputs being shown by the animal, as when a gull's state is judged as "nesting-building" during the spring period when it is collecting plant material, assembling the material in specific sites, *etc.* Any indication the ethologist can devise that reflects the internal state of behavioral readiness of an animal usually leads to higher predictability concerning its behavioral outputs. So long as one articulates the list of alternative states, *table 2-II* may be used to analyze the behavioral data.

When other than transitional criteria are used to judge the behavioral state, *table 2-II* is no longer a transitional matrix of behavior. It cannot be used, for example, to deduce a sequence of spontaneous behavior under constant conditions of the external environment. For this reason, *table 2-II* was labeled a *state matrix* to reflect its general nature. As a non-transitional state matrix, the columns of *table 2-II* (p. 35) constitute a list of *behavioral vectors* of probability, denoted β_j. That is, each column is a vector whose members are the probabilities $p(b_i|\beta_j) = p(i|j)$ and $\sum_i p(i|j) = 1$.

inputs and control

The immediate control of behavior in an individual is dictated by internal and external factors; logically there is no third possibility. Let each discriminable combination of external factors be called the *external input*, denoted E_k ($k = 1, 2, 3, \ldots, m$). Each input specifies a different state matrix like that of *table 2-II*, and in so doing specifies the behavioral vector of outputs according to the state of the animal at the time of receipt of the input. One way to visualize the conceptual scheme is via a *control matrix* of internal behavioral states and external sensory inputs, which together specify a particular probabilistic distribution or vector of behavioral probabilities (*table 2-III*). Or, if the composition of the vectors is to be included, the control matrix has a third dimension

Table 2-III

Control matrix of behavioral vectors.

		external input (E_k)				
		1	2	...	k	...	m
	1	B_{11}	B_{12}	...	B_{1k}	...	B_{1m}
	2	B_{21}	B_{22}	...	B_{2k}	...	B_{2m}
behavioral state (β_j)	...	...	...	...	...	...	...
	j	B_{j1}	B_{j2}	...	B_{jk}	...	B_{jm}
	...	...	...	...	...	...	...
	ν	$B_{\nu 1}$	$B_{\nu 2}$	...	$B_{\nu k}$	...	$B_{\nu m}$

of outputs and its cells are the probabilities $p(b_i|\beta_j E_k)$ $\equiv p(b_{i|jk}) \equiv p(i|jk)$.

In order to make the notion of behavioral control precise, it may be expressed as a *behavioral control function* (*C*), which specifies a mapping of each pair of behavioral states (β_j) and external inputs (E_k) onto a behavioral vector (B_{jk}) of probabilities of outputs. Symbolically,

$$C: \quad (\beta_j, E_k) \mapsto B_{jk} , \qquad\qquad (2.7)$$

where $B_{jk}: p(b_{i|jk})$.

Although the notation of *eq (2.7)* is necessarily involved for logical clarity, the idea of behavioral control functions is a straightforward one. For example, Beer (1961, 1962) distinguished outputs of the black-headed gull on the nest: $b_1 \equiv$ resettling with quivering, and $b_2 \equiv$ resettling without quivering. Two behavioral states were distinguished by the general ongoing behavior of the birds: $\beta_1 \equiv$ egg-laying state, and $\beta_2 \equiv$ incubating state. Finally, two external inputs were distinguished: $E_1 \equiv$ one egg in

the nest, and $E_2 \equiv$ three eggs (the normal clutch size).
Beer created three-egg clutches during laying by adding
two eggs to a nest in which the first egg had been laid
and created one-egg clutches during incubation by removing two of the eggs from the nest. The probability of
b_1--and hence the probability of b_2, which in this case
is $1-p_1$--was different in all four combinations of input
and state. Whether or not a gull would quiver in a particular instance of resettling was never predictable except as a probabilistic statement, which is to say a behavioral vector B_{jk}.

These external sensory inputs have high diversity.
There are so many distinguishable alternative stimulus
patterns that can fall upon the retina that one almost
believes visual inputs alone to be infinite in number.
However, there are certainly limitations on how many inputs the visual system can distinguish as different, and
some attempts have been made to determine the entropic
rate of input accepted by human subjects viewing random-
dot patterns in quick succession. If such experiments
were extended to include the color of dots, acoustic inputs, chemical inputs, *etc.*, the average rate of sensory
input would be enormous. Consider a vertebrate eye with
more than 10^8 receptor cells. A fanciful calculation
that ignores color vision, considers each cell merely to
be on or off (no gradations) and considers all spatial
patterns created by the retinal stimulation to be distinguishable yields an *input entropy* of $'H = \log_2 10^8 =$
26.6 bits/input pattern. Distinguishing a pattern each
second provides an input entropic rate of more than 25
bits/s.

behavior

During the foregoing development of behavioral control, it was noted that the maximum output entropy of behavior is on the order of 10 bits/s (probably *much* less.),
whereas the input entropy might be on the order of twice
that value (probably *much* more). Similar fanciful calculations may be made concerning the energetic exchange during behavior.

Output energy may be calculated from metabolic rates
by assuming a homeotherm burns 5 Calories of carbohydrates
per liter of oxygen consumed. A hummingbird at rest thus

expends about 10^4 erg/s, in flight about 10^5, and a resting cow expends about 10^7 erg/s. Sensory inputs, however,
have little energetic content. For example, a photon of
light *(ch 3)* at 500 nm of wavelength carries about 4×10^{-24}
erg. If three photons are necessary to activate a given
photoreceptor sufficiently to cause visual sensation, the
energetic input is about 10^{-23} erg/cell, and if half the
eye's 10^8 cells are activated during every second, the average requirement is 10^{-19} erg/activation. If activation
lasts for only 1/1000 s, the eye still requires input energy of only about 10^{-16} erg/s. The qualitative point is
simply that input sensory energy is far less than output
energy expended, even while resting. Sensory energy is
not (and could not be) the source of energy expended in
behavioral outputs: that source is food.

These comparisons lead to a very general statement
about behavior. *Behavior* is characterized by entropic and
energetic transductions by an organism, in which the long-
term averages convert high entropic and low energetic sensory inputs into low entropic and high energetic outputs
(fig 2-4). I believe that Keith Nelson pointed this out
to me years ago, but only recently have I recognized the
implication of this relation for entropic exchange during
animal communication, discussed in the next major section.

*Fig 2-4. Characteristics of behavior include the trans-
duction of high entropic, low energetic inputs into low
entropic, high energetic outputs over long-term averages.*

Communicative Behavior

It is now possible to combine the notions of signals,
information and behavior into a framework of communicative
behavior. Like other complex phenomena, communicative behavior has certain emergent properties that come from the
interactions of the three elements mentioned and could not
have been predicted by considering the elements separately.
Figures 2-1 and *2-4* together suggest that signals result

from the sender's behavioral outputs, which signals in
turn act as external inputs to the receiver. The aim of
this section is to scrutinize the implications of this
relationship.

outputs as signals as inputs

In generating an output (b) the animal also generates
physical disturbances (potential signals). The animal expends considerable energy in generating outputs, but little
of this energy goes into the physical disturbances that
are potential signals. Some energy is lost in the generating process according to the second law of thermodynam-
ics, other energy is transferred to the environment in ways
largely irrelevant to communication, some energy is exchanged in form, and a very small amount of expended energy
may be transmitted by the signal. For example, one human
output is waving. In raising the arm some energy is lost
in the process as heat, some energy is transferred to the
environment in the form of kinetic displacement of air molecules, some energy is exchanged for potential energy (and
hence regainable as kinetic energy when the arm falls), but
little of the animal's energy becomes part of the optical
signal of reflected sunlight.

Outputs have diversity, and hence calculable output
entropies H'. The maximum output entropy is low, but the
maximum entropy of signals (H^*) generated by the outputs
is lower ($H'_{max} \geq H^*_{max}$), although this point is not immediately obvious. Outputs are distinguished from one another
by differences in the signals they generate, so the set of
all signals generated must be a subset of the set of all
outputs. If this set-relation were found untrue, certain
distinguishable outputs would be logically unrecognized,
and the output set would have to be enlarged by further
distinctions among outputs.

Every received signal is an external input, but not
the reverse, so that signals are a subset of sensory inputs. This means that physical disturbances have no role
in communication unless they specify state matrices, which
is to say they combine with behavioral states to dictate
behavioral vectors of output probabilities according to eq
(2.7). But the set of all inputs and the included subset
of signal-inputs have diversity, in the sense of having
two or more members, each with an associated probability.

If $^*H$ be the entropy of signal-inputs, it follows from *eq (2.1)* and the set-relationships that $'H_{max} \geq {}^*H_{max}$ in bits/input, and hence also in bits/s.

There comes a point in the construction of broad models of living systems that deduction must give way to reasonable assumptions, based if possible on empirical generalizations. Since such assumptions should be made explicit, I ask acceptance of the reasonableness that long-term entropic averages follow the same relations as maximum entropies; *i.e.*, that

$$\hat{H}' \geq \hat{H}^* \text{ , and} \tag{2.8}$$

$$'\hat{H} \geq {}^*\hat{H} \text{ .} \tag{2.9}$$

context

It was established in the previous major section (see *fig 2-4*) that input entropies exceed output entropies in the maximum and on the long-term average. This is a necessary deduction from principles of control theory (Ashby, 1961), since if inputs are to control outputs they must be just as diverse:

$$'\hat{H} \geq \hat{H}' \text{ .} \tag{2.10}$$

Combining this inequality with those of *eqs (2.8)* and *(2.9)* leads to the statement

$$'\hat{H} \geq {}^*\hat{H} \geq \hat{H}' \geq \hat{H}^* \text{ .} \tag{2.11}$$

This relation presents no real difficulties for behavioral transduction because input and output entropies can be decomposed into signal and non-signal components. Let non-signal components be denoted ^OH and H^O; then *fig 2-4* may be revised as shown in *fig 2-5*.

Equation (2.11) does, however, present an apparent paradox for communicative behavior. If the communicants are similar, in the sense that two members of the same species are similar, they have similar entropic relations. However, *eq (2.11)* shows that the output signal-entropy of the sender ($^*H$) is at best equal to, but usually *much* less than, the input entropy of the receiver ($'H$). In short, signals themselves are not sufficiently diverse to control the communicative behavior of receivers.

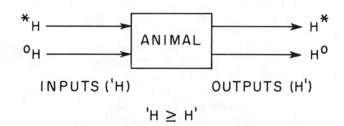

INPUTS ('H)   OUTPUTS (H')

$$'H \geq H'$$

Fig 2-5. Partition of entropies of fig 2-4, showing that input entropy ('H) may be decomposed into entropy from signals of other animals (*H) and entropy due to non-signal sources collectively called context (^{O}H). Similarly, output entropy is divided between signals (H*) and outputs that are not signals (H^{O}).

The point may be made in another way by considering two communicants, A and B. The signal-output of A (H^*_A) is at best equal to the signal-input of B $(*H_B)$, and usually will be much less because B will not receive or distinguish all the signals of A. Then B's signal-output entropy (H^*_B) will be yet smaller, but by *eq (2.11)* should exceed A's output-entropy if B's signals are input to A. Then A's output-entropy (H'_A) is larger than A's signal-entropy (H^*_A), so that one has come full circle. The paradox is thus that the H^*_A of the first exchange of signals is shown to be larger than the H^*_A of the next exchange. Were this to be true, entropies would run downhill quickly during life.

The apparent paradox is resolved by recognizing that each communicant must be receiving additional entropy from sources other than signals (*i.e.*, ^{O}H), as shown in *fig 2-6*. The set of all such non-signal inputs may be called the *context* of communication, the importance of which has been usefully emphasized by W.J. Smith (*e.g.*, 1968).

the "CB" model

This is a favorable juncture for an aside concerning a commonly used ethological model of communication that might be confused with the notions developed in this chapter. This model treats communication as if the animals exchanged signals in some discrete, alternating fashion.

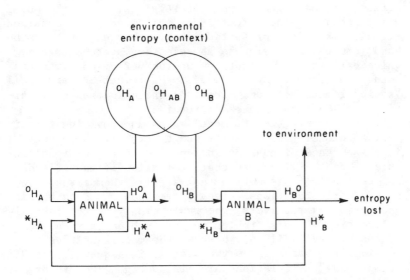

Fig 2-6. Entropic flow in communication, combining notions of figs 2-1 and 2-5. The Venn diagram at the top indicates the partial overlap of contextual sources of input to two communicants, and what appears to be the same signal (e.g., H_A^* and $*H_B$) is given two denotations because the signal can be transformed in passage by noise (see fig 2-7, below).

For example, A signals to B, then B to A, then A to B, and so on. I call this the "citizen-band" (CB) model of communication, in analogy with citizen-band radio in which the sender concludes his broadcast with "over," listens to the other communicant until she says "over," and then sends a-gain.

Studies utilizing this CB model construct a transitional matrix of A's signals and B's signals, and then perform Markovian analyses using equations similar to those presented in this chapter. The CB model therefore resembles in certain formal and quantitative aspects some of the elements of the conception of communication developed here.

The CB model is, however, quite different. Also, it makes several hidden assumptions about the communicative process. Schleidt (1973) has thoroughly criticized the CB model, and it is necessary only to add a few notes here. As Schleidt points out, the model cannot handle long-term

effects of the signals upon the receiver, since the model
assays only for the effects upon the immediately returned
signal. The model presented in this chapter, on the other
hand, conceives of signals as selecting state matrices,
which carries no implications whatsoever of an immediate
change in output. My model allows for what Schleidt calls
"tonic effects" of communication: the maintenance of particular behavioral states in receivers.

Furthermore, the CB model makes no provision for either spontaneity of behavior due to transitions of outputs
in the absence of changes in input, or for non-signal inputs of context. Yet another limitation of the CB model
is its implicit assumption that signals themselves have
no syntactic relations: signals are treated as being read
one-by-one by receivers, who reply simply to the last signal received. This problem merges into another one pointed out by Baylis (1976): because the model excludes the
possibility that A gives two signals in sequence, the "informational" calculations are misleading. The "information"
calculated by the CB model, by the way, does not measure
the amount of information being transferred between communicants; at best it measures what the observer learns
by watching the communicative interaction. Even this calculation is open to doubt, and Baylis has provided an alternative method for calculating the total amount of information about the system that one obtains by observing
it.

Basically, the criticisms amount to this. The CB model assumes that A signals, B replies, A signals again, and
so on. Only in very special circumstances, if ever, is animal communication so simply described, and even when it
happens in this way the model for treating it ignores valuable data. The purpose here, however, is not to scrutinize the inadequacies of the CB model, but rather to draw
attention to its existence so that it will not be confused
with the more general conception of communication presented
in this chapter.

feedback

When the present value of a variable influences its
subsequent value the system containing the variable exhib-
its feedback. There are three principal sources of feedback by which an animal's output influences subsequent
outputs. One of these is represented by the state matrix

of *table 2-II* (p. 35): the present state (β) of an animal
determines stochastically its next output (b), which in
turn may affect markedly the behavioral state, particularly in semi-Markovian behavior.

The other two routes of feedback in a simple communication system are external. The output-signals of one
animal become the input-signals of another, thus affecting
its output according to *eq (2.7)*. The output-signals generated by this second animal can in turn act as inputs to
the first, thus closing the feedback loop of reciprocal
communication.

The final feedback is via context provided by the environment. Non-signal outputs of an animal may affect the
environment (some of the output entropy is lost); in effect
these outputs act as signal to components of the environment when the components are treated as receivers. In dealing with animal communication *(figs 2-2* and *2-3)* it is convenient to consider only sender's outputs that are directly
inputs to another animal as the "signals." The environment, in turn, provides as its outputs variables that act
as inputs to animals, once again closing a feedback loop.

Some ethologists write as if the contextual entropy
were shared by the communicants, but this is only partly
true. No two animals have precisely the same external inputs, so their contexts will always differ at least slightly. This fact is incorporated into *fig 2-6* (p. 45), which
diagrams the external feedback loops of entropic flow in
a simple communication system.

recognizing communication

The definition of communication as a transfer of information via signals in a channel between sender and receiver still serves well after the development and scrutiny of a broad model of animal communication. The definition does not, however, instruct the observer as to
how to recognize communication. The operational instruction of Klopfer and Hatch (1968: 32) has been accepted by
others (*e.g.*, Schleidt, 1973: 359): "the ultimate criterion for recognition ... is that of a resultant change,
sometimes delayed, sometimes scarcely perceptible, in the
probability of subsequent behavior of the other communicant." I should like to offer minor improvements in the
instruction. Conceive of two behavioral situations identical in external input (E_k) and behavioral state (β_j) of

the reputed receiver, but differing in the fact that the
external input of one includes a reputed signal absent in
the other case. If the behavioral vector (B_{jk}) of the reputed receiver differs in the two cases, I shall say that
communication has occurred: there has been a transfer of
information due to the signal. This is a more precise
rendering of the instruction in Hailman (1977a).

It is useful to conclude the discussion of basic
communicative behavior with an example that emphasizes the
irrelevancy of immediate changes in receiver-output as the
criterion for recognizing communication. Suppose a cat is
crouched at a mouse-hole. Under constant external conditions, the cat may remain crouched for some time, but as
the continual output of crouching alters its internal behavioral state, the cat will eventually "give up" and
leave the hole. This change in output is spontaneous, and
not well labeled as a "response." Employing the operational instruction for recognizing communication, consider a
similar incident in which just before the cat would have
left, a mouse in the hole ventures sufficiently far up to
be seen by the cat. Instead of leaving, the cat now remains at the hole. However unfortunate for it, the mouse
has communicated with the cat, and we as observers have
recognized the communication--not as a change in output
of the cat (indeed, the effect was to prevent a change in
output)--but rather as a redistribution of probabilities
of output: a change in behavioral vector.

Final Notes on Communication

Semiotics, cybernetics and ethology provide a nearly
endless array of issues about communication, any one of
which could probably occupy an entire volume. The purpose
of this final section is to note a few communication topics
and phenomena used in analyses of optical signals in later
chapters.

semantics, syntactics and pragmatics

Peirce's ideas of semiotics were extended by Morris
(e.g., 1946) and Cherry (1957) in various ways. Morris
appears to have identified the triad of semiotic problems:
semantics, syntactics and pragmatics. For purposes of the
restricted aims of this volume, semantics may be taken as

---

the study of relations of signals to their referents, *syntactics* the study of the relations among signals, and *pragmatics* the study of the relations of signals to their effects. Weaver, in an insightful essay appended to Shannon's development (Shannon and Weaver, 1949), rephrased the semiotic problems, which he called levels of analysis. Weaver asked how precisely transmitted signals conveyed the sender's intended meaning (semantics), how accurately signals were transmitted (syntactics) and how effectively the received signals affected the receiver's conduct (pragmatics). Cherry's (1957) treatment also differed subtly from Morris's, and appears to have been the primary influence on ethologists who have used the framework (*e.g.*, Marler, 1961; Smith, 1968). Smith is concerned principally with the factors that produce signals, which he calls the "messages" of communication (semantics) and the effects of signals on the behavior of receivers, which he calls the "meanings" of communication (pragmatics). My purpose is not to explore the differences in usages of the semiotic triad, but merely to use a version of it to organize factors relating the kind of information transferred to the design of optical signals *(ch 8)*.

Peirce himself provided the initial classification of semantic relations of signs to their objects, which I use in the physical terms of signals and their referents (pp. 26-27). He divided signs into *indexes*, which point out their objects; *icons*, which resemble theirs; and *symbols*, which stand for theirs in other ways. This traditional classification proves useful in analyzing semantic characteristics of optical signals in *ch 8*.

deception, distortion and noise

One sometimes thinks of communication as a simple process by which the sender informs the receiver about himself or other phenomena in the world, but the process is plagued with nuances that render it anything but simple. For example, the sender may transmit "misinformation" by promulgating a signal that the receiver interprets as standing for something that it does not. *Chapter 6* deals with the problem of *deception* in terms of interspecific communication as a prelude for considering forms of interspecific deception in *ch 8*.

Signals as received are not always identical with signals sent. The simplest kind of change in a physical signal during transmission is *distortion*, in which the

change entails no loss of information transferred. Temperature differences or salinity differences in water may
create optical lenses that distort the perceived shapes
of fishes and other animals, but the SCUBA diver and probable the animals themselves can often still identify the
species-specific characteristics that serve as optical
signals.

When the informational content of the signal sent is
altered during its transmission, the disturbance is called
noise. The notion of noise comes directly from static in
radio transmission, where one may mistake parts of the
signal sent for something different because of sounds added
or changed by electrical interference. *Chapter 7* considers
kinds of optical noise and their effects on the design of
signals.

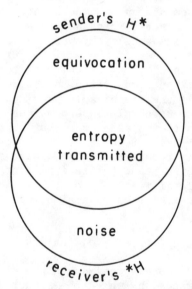

*Fig 2-7. Relation between
noise and equivocation,
defined in a Venn diagram
showing the entropic con-
tent of the signal as
sent (H*) and as re-
ceived (*H) in a noisy
channel. The intersec-
tion is the "useful" in-
formation communicated.*

Noise is measured by the confusion it causes in the
receiver, the quantity Shannon named *equivocation*. Equivocation is that part of the sender's signal-entropy that
is not available to the receiver, and the set-relationships of noise and equivocation may be diagramed as in
fig 2-7. In essence, noise is entropy received as signal
but generated by some source that the receiver cannot dissociate from the sender.

---

redundancy

Perhaps Shannon's most startling discovery about communication (Shannon and Weaver, 1949) is that it is always
possible to transmit error-free signals in a noisy channel
at the sacrifice of delay in transmission time. The technical details are not relevant to the present study, but
an important point in Shannon's development is that redundancy may be used to combat equivocation. One often
thinks of redundancy as simple repetition, but technically
redundancy is the degree to which a channel is not efficiently utilized to transmit information. Let H_{obs} be the
observed or utilized entropy and H_{pos} be the possible entropy that could, with efficient coding, be transmitted.
Then the redundancy is

$$R = (H_{pos} - H_{obs})/H_{pos} . \qquad (2.12)$$

One kind of redundancy is serial redundancy, in which
the occurrence of some signal in sequence is at least
stochastically predictable from preceding signals. (Behavioral outputs are quite serially redundant, a fact that
allows the sequential analysis by Markov chains discussed
earlier.) An extreme case of serial redundancy is the letter following "q" in English words.

That redundancy may be used to combat equivocation
due to noise may be illustrated by a simple example of
serial redundancy. The word "rite" might be disrupted
during telegraphic transmission to be received as "@ite,"
in which case the receiver must try to decide whether the
word sent was *bite, cite, kite, mite, rite, site,* or even
colloquial words such as *fite, hite, lite, nite* or *tite.*
If the received word were "r@te," the sent word might be
rate, rete, rite or *rote.* On the other hand, if "quite"
were disrupted to "@uite" it is likely to be only *quite*
or *suite;* if disrupted to "q@ite" it is almost certainly
quite. Total serial redundancy is not necessary to a-
chieve protection against the effects of noise. In general, English is highly redundant at many levels, and the
longer the word, the less likely will disruptions in its
transmission prevent the reader from correctly interpret-
ing it. That English is redundant is obvious from the
fact that not all combinations of letters are words *(eq
2.12).* As a result, we are likely to recognize *teh, het,
hte* and other typographical slips as "the."

The calculation of redundancy in animal signaling systems is usually difficult. Rand and Williams (1970) report eight species of anoles known or suspected to occur at La Palma, Hispaniola. The males of each species are differently marked, but they differ in more ways than necessary to distinguish them. In dewlap alone there are four sizes, 11 colors and two patterns giving a total of 924 possible combinations (depending upon what one considers "possible" in real biological terms). Therefore, if the species are equally abundant, H_{pos} = $\log_2$ 924 = 9.85 bits/dewlap and H_{obs} = $\log_2$ 8 = 3 bits/dewlap, so that by *eq (2.12)*, R = 0.7 or 70% redundancy. This figure is the redundancy to the scientist identifying anoles at random; it may be more difficult to calculate a meaningful redundancy from a given anole's viewpoint of merely distinguishing its own species from all others combined. Although the ultimate understanding of animal communication must deal with all problems of redundancy, the only form of redundancy relevant to the present level of analysis of optical signals is simple repetition *(chs 6 through 8)*.

Overview

Communication is the transfer of information via signals sent in a channel between a sender and a receiver. The occurrence of communication is recognized by a difference in the behavior of the reputed receiver in two situations that differ only in the presence or absence of the reputed signal. The behavioral difference, however, is not necessarily a simple change in what the animal does (behavioral output) just after receiving the signal; rather, the signal causes a redistribution of the probabilities of outputs (a change in behavioral vector). Therefore, the effect of a signal may be to prevent a change in the receiver's output, or to maintain a specific internal behavioral state of readiness. The understanding of communicative behavior requires a synthesis of semiotics, cybernetics and ethology; a preliminary framework is offered. The analysis of optical signals uses the basic framework to characterize the communicative system: the channel is considered first *(ch 3)*, then the sender *(ch 4)* and the receiver *(ch 5)*. Deception is the transfer of "misinformation," an important aspect of interspecific communication *(ch 6)* that has later relevance to social

signals as well. Noise is information added to signal-
information in such a way that the receiver cannot dissociate the two, and hence suffers confusion (equivocation). Optical noise plays an important role in visual
signaling *(ch 7)*. Finally, the problems of what kind of
information is transferred and how it is encoded involves
semiotic notions of semantics, syntactics and pragmatics,
considered in relation to optical signal-design in *ch 8*.

Recommended Reading and Reference

Sebeok's (1968, 1977) volumes offer the best introduction to various views attempting to synthesize semiotics,
cybernetics and ethology; Johnston's (1976) recent paper
is an important adjunct. The entire September 1972 issue
of *Scientific American* is devoted to communication in its
various ramifications.

C.S. Peirce's semiotic ideas may be found in a collection edited by Hartshorne and Weisee (1931-35) and in
his letters to Lady Welby (Lieb, 1953). Influential derivative works include Morris (1946) and Cherry (1957),
and important contributions from other viewpoints include
Chomsky (1957) and Pierce (1961). The foundations of cybernetics are Wiener (1948) and Shannon and Weaver (1949),
but Ashby's (1961) treatment of control systems is independent of linear systems of engineering, and Singh's (1966)
work on information theory is a fine introduction. Attneave
(1959) is another useful introduction to information theory,
especially for behavioral scientists.

Tinbergen's (1951) *Study of Instinct* is a classic of
ethology, now seriously outdated, but a thoroughly operational and comprehensive replacement has yet to be written.
Hinde (1970) is a useful introduction to problems of control and ontogeny, and Brown (1975) to evolutionary questions about behavior. The interdisciplinary elements that
contributed to modern synthetic ethology may be appreciated
through Klopfer's (1974) historical treatment.

Chapter 3

THE CHANNEL

3

God said, "Let Newton be!" and all was light. --Alexander Pope.*

In order to understand the optical design of animal signals one must study how they are generated, transmitted and received. This chapter is the first of three providing relevant background. It deals with physical properties of light, to be followed by chapters dealing with generation and reception of optical signals.

The choice of what constitutes the channel between sender and receiver *(fig 2-1)* is somewhat arbitrary. One might consider the entire informational processing line between the central nervous system of the sender and that of the receiver as the channel. In fact, Shannon (Shannon and Weaver, 1949) and others conceive of a communication system as consisting of an informational source connected to a transmitter that generates signals on one end of the channel and a receiver *per se* connected to a destination at the other end. For present purposes,

---

*Pope's "triple-entendre" comes from *Epigram on Sir Isaac Newton*: "Nature and Nature's laws lay hid in night: God said, 'Let Newton be!' and all was light." J.C. Squire (1884-1958) replied "It did not last: the Devil howling 'Ho! Let Einstein be!' restored the status quo." Relativistic considerations are omitted from this chapter.

boundaries are conveniently set at the physical limits of
the communicating organisms: one is a sender, the other a
receiver in the broad sense, and everything in between is
the channel.

In animal communication, the optical channel consists
of electromagnetic energy propagated through air or water.
In order to understand this channel it is necessary to review some relevant facts about the physics of light, ambient radiation from the chief source of light (the sun)
and the propagation of light through media that may contain suspended materials (e.g., water vapor in air) and
macro-objects (e.g., plants). All these topics are used
in later chapters in attempting to uncover the design of
optical signals used in animal communication.

Light and Information

In order to transfer information, signals must have
variety (ch 2). Therefore, understanding of the optical
channel begins with sources of variety in signals: the
physical variables of light.

the photon

Light is electromagnetic radiation that causes visual sensation. Electromagnetic radiation is emitted and
absorbed in indivisible units called *quanta*, a quantum of
light also being known as a *photon*. The photon may be
visualized as a moving wave having electric and magnetic
vectors (fig 3-1). Each quantum has a measurable *wavelength* (λ) and *speed* of movement (c), related by the general wave equation

$$c = \lambda \nu \, , \tag{3.1}$$

where ν is the *frequency*. Quanta having frequencies in
the range of $750 \geq \nu \geq 400$ THz are absorbed by the human
photoreceptors and hence define the *visible spectrum*. The
energy of a photon is proportional to its frequency:

$$E_q = h\nu \, , \tag{3.2}$$

where the proportionality constant h is Planck's constant

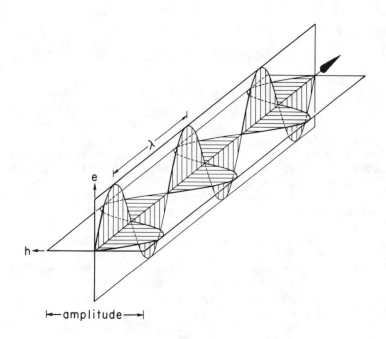

Fig 3-1. Representation of a photon moving from lower left to upper right with its electric vector (e) shown in the vertical plane and its magnetic vector (h) at right angle and in phase. The wavelength (λ) is shown as the distance between peaks, and the number of peaks that pass a point in space in a given time is the frequency (ν), related by the photon's speed (eq 3.1).

$(6.624 \times 10^{-27}$ erg-s).* The speed of light in a vacuum (c) is about 3×10^8 m/s, so by *eq (3.1)* the wavelengths of visible light are $400 \leq \lambda \leq 750$ nm. Other media slow light to some value $\overline{c}^{\iota}$, and the ratio

$$n = c/c^{\iota} \, , \qquad\qquad (3.3)$$

is the *index of refraction.* When a photon is slowed its frequency is invariant, but its wavelength changes accord-

---

*The magnetic vector is also denoted by *h*, but that symbol is used only in *fig 3-1* above.

ing to *eq (3.1)*.

light

 A few photons absorbed by the eye in quick succession are sufficient for the minimum visual excitation, but ordinarily the temporal flux is very high before we are aware of light. The flux is expressed by the number of quanta per unit time or the amount of energy per unit time (related by the frequency according to *eq 3.2*). The latter is called *radiant power* (measured in watts, where $1 W = 1$ Joule/s and $1 J = 10^7$ erg). *Radiance* is the power emitted from a radiant area such as the sun's disk and *irradiance* is the power falling upon an irradiated area such as an animal's body; both are expressed in units such as W/m^2 and denoted I, but context usually distinguishes them.

 Light has emergent properties due to its composition of different kinds of photons. When all are of the same frequency (wavelength), the light is *monochromatic*, and if their waves are exactly in phase the light is *coherent*, as from lasers *(fig 3-2)*. Spectrally *complex* (*i.e.*, not monochromatic) light may be characterized by a density-distribution of irradiance per spectral increment through the spectrum (*e.g.*, $W m^{-2} THz^{-1}$ as a function of THz). When the electric (and hence also magnetic) vectors of the photons are aligned the light is *plane-polarized (fig 3-3, p. 60)*, and there also exist more complicated kinds of polarization.

 One may imagine a spatial array of photons traveling toward an observer in which the component parts of the array may differ in spectral-power distribution and degree of polarization. The possible compositions of such an array are effectively without bound, and the rapidity with which the composition can change in time is likewise nearly limitless. Therefore, the capacity of the optical channel to carry information *(ch 2)* is virtually infinite: a given array has an indefinitely large number of alternative states, and these states can change very rapidly.

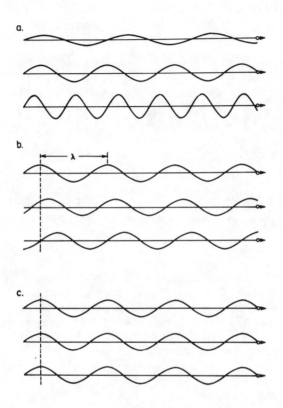

Fig 3-2. Complex, monochromatic and coherent light shown by electric vectors of photons moving from left to right. At top (a) the three photons differ in frequency, hence their wavelengths (eq 3.1) also differ. The group constitutes spectrally complex light. At middle (b) the three photons are of the same frequency (wavelength), but out of phase, and hence constitute monochromatic light. At bottom (c) they are in phase, so the light is also coherent.

Light and Matter

If there are bounds to the capacity with which information is transferred by light, these bounds relate to the sender's ability to promulgate signals or the receiver's ability to accept them, or both. As necessary background to these limitations, to be considered in the following two chapters, one must understand how light interacts with matter.

UNPOLARIZED

VERTICALLY
PLANE-POLARIZED

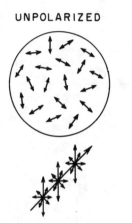
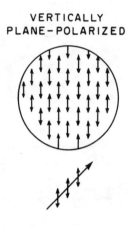

Fig 3-3. Plane-polarization of light diagramed as electric vectors seen from in front of on-coming photons (upper diagrams) and as commonly represented (lower diagrams). In ordinary unpolarized light (left) the planes are not aligned, whereas in plane-polarized light (right) all electric vectors are in parallel planes.

absorption and emission

Energy and mass are convertible according to Einstein's famous equation, but in ordinary physical interactions on earth energy is conserved during transitions from one form to another. An atom or molecule of matter *absorbs* a photon by converting its electromagnetic energy to electronic energy, which raises the molecule from a ground state to an excited state. Spontaneous return to the ground state *emits* a photon, and in both absorption and emission the difference in energy between ground and excited states is exactly the energy of the photon *(eq 3.2)*. *Induced emission* occurs when one quantum absorbed by an excited molecule causes it to emit multiple quanta in the return to the ground state.

Single atoms and simple molecules have only a few possible ground and excited states, whereas complex molecules have many possible states. Hence the former can undergo limited electronic transitions, which absorb or emit specific frequencies according to *eq (3.2)*. These simple molecules exhibit *line spectra* of emission or absorption such as those used by astronomers to identify the chemical composition of stars from their light. Complex molecules, on the other hand, have ever-changing states described by density-distributions of probability as a function of trans-ition-energy. Therefore, they show *continuous spectra*: probability of absorption or emission as a function of frequency, or (over some finite time-increment) power ab-sorbed or emitted as a function of frequency. For this reason, absorption spectra of the complex visual-pigment molecules in the eye resemble Gaussian, bell-shaped proba-bility density-distributions (*ch 5*, below).

Energetic relations in molecules concern heat and other forms of energy apart from photons. Adding energy to substances, such as electric current passed through a tung-sten wire, heats the molecules to *incandescence* so that they emit energy in the form of photons (thermoluminescence). An ideal incandescent radiator (the *black body*) emits a spectrum that depends upon its absolute temperature accord-ing to an equation known as Planck's law. Conversely, ab-sorbed photons may be converted to heat and other internal energy, causing degeneration of the excited state without emission of light.

A particular kind of emission occurs when a quantum is absorbed and the molecule exchanges energy internally while in the excited state (relaxation). The molecule thus de-generates to a lower excited state and then emits a quantum of energy less than that absorbed. The emitted quantum must therefore have a lower frequency than the absorbed quantum (*eq 3.2*), the familiar example being visible *fluor-escence* of light from a substance irradiated by higher-frequency ultraviolet radiation. When fluorescence is de-layed because of certain electronic structure of the mole-cules involved, the phenomenon is *phosphorescence*.

Other forms of light-emission always involve adding energy by some mechanism, which may be quite complex in some cases. Fusion and fission processes, such as those in military bombs and in the interiors of stars, may provide the energy. Modern technology uses electric current applied to thin conducting panels to create *electroluminescence*, and chemical reactions may provide *chemiluminescence*. Cer-tain animals, such as fireflies, use chemical reactions to create *bioluminescent* signals (*ch 4*).

---

reflection and transmission

Most animal signals are created by reflecting ambient
light from some surface, such as the animal's body, which
becomes a *secondary source* of radiation. In a gross sense,
light striking a non-fluorescing substance has three fates:

$$I_0 = I_r + I_a + I_t \, , \tag{3.4}$$

where I_0 is the incident radiation, I_r the radiation reflected, I_a the radiation absorbed and I_t the radiation
transmitted *(fig 3-4)*. When $I_t \doteq I_0$, the substance is

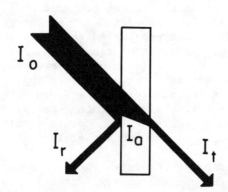

Fig 3-4. *The fate of
incident light (I_0)
striking a substance.
Some light may be re-
flected (I_r), other
absorbed (I_a) and the
remaining transmitted
(I_t) through the sub-
stance.*

transparent and when $I_t \doteq 0$, it is *opaque.* All substances
absorb at least some of the incident radiation (often turning it to heat), and most either reflect or transmit the
rest, although some substances do all three. *Equation (3.4)*
may have different values at different frequencies, so it
becomes useful to define *coefficients* as $r = I_r/I_0$, $a =
I_a/I_0$ and $t = I_t/I_0$, and rewrite *eq (3.4)* as

$$I_0 = rI_0 + aI_0 + tI_0 \, , \tag{3.4a}$$

where $1 = r + a + t$. (A coefficient multiplied by 100 is
the *percent absorption, reflection* or *transmission*: $A = 100a$,
$R = 100r$ and $T = 100t$.) The coefficients may vary with
frequency, and hence a plot of r *vs* ν (or λ) is the *reflectance spectrum* and of t *vs* ν the *transmittance spectrum*
of the substance. Most substances that both reflect and

transmit have similar reflectance and transmittance spectra, but these two spectra may differ (*e.g.*, in gold leaf), a phenomenon known as *selective reflectance.*

When the value of a is independent of frequency, the substance is spectrally *neutral*, as with neutral density filters. In this case it is convenient to define *density* as

$$OD = -\log_{10} t \;, \tag{3.5}$$

where OD is also known as *optical density* (see next section, below). Densities conveniently add, so that a filter transmitting 0.1 I_0 (density 1) in series with a similar filter has a combined density of 1+1 = 2 and a combined transmission of 0.1 × 0.1 = 0.01 or 10^{-2}.

absorption, extinction and optical density

The transmission of light through a substance depends upon the spectrally dependent *extinction coefficient* (α) of the substance and the distance (d) that light travels through it:

$$I_t = I_0 \varepsilon^{-\alpha d},$$

where ε is the base of the natural logarithm. (For chemical solutions, d includes a factor for the concentration of solute as well as the length of light-path.) Because $t = I_t/I_0$, it follows with rearrangement and passing to logarithms that

$$-\log_\varepsilon t = \alpha d \;.$$

When common logarithms are used instead of natural logs, the quantity is *optical density* (*eq 3.5*). Normalized optical densities are independent of the thickness or concentration of the substance because distance cancels from ratios such as $\alpha d/\alpha_{max} d$. Therefore, each substance has one *relative extinction spectrum*, but an indefinitely large number of absorption spectra, depending upon the thickness (or concentration).

Absorption (a) and extinction (α) coefficients are related by optical density. By *eq (3.5)*, $OD = -\log t$, so that $t = 10^{-OD}$. Furthermore, assuming negligible reflectance, $t = 1-a$ by *eq (3.4a)*, so that

$$a = 1 - 10^{-OD}.$$

This equation may be expanded to an infinite series (Dartnall, 1957), the first term of which is about $2.30D$, succeeding terms being fractions whose numerators are increasing powers of $2.30D$. If the light-path is very short (or the solution is very dilute), so that optical densities are small ($OD < 0.05$), the numerators of the series are fractions themselves, yielding very small second and subsequent terms, which may be disregarded. Therefore,

$$a \doteq 2.30D \, ,$$

and since the light-path or concentration is also negligible, absorption is nearly proportional to extinction. For this reason, absorption spectra of very dilute solutions are treated as if they were extinction (optical density) spectra, a fact utilized in characterizing of visual-pigment absorption *(ch 5)*.

reflection, refraction and polarization

On a microscopic level the angle of incident light equals the angle of reflected light. Therefore, on a micro-smooth surface, most of the incident light reflects at one angle (*specular reflectance*) that is equal to the incident angle *(fig 3-5a)*. Fish scales, for example, reflect specularly *(ch 4)*. On a micro-rough surface, light is multiply reflected at the surface before being propagated into space, and hence emerges in various directions (*diffuse reflectance, fig 3-5b*). Most bird feathers, for example, reflect diffusely. An ideal diffusing surface obeys *Lambert's law*:

$$I_\phi \propto \cos \phi \, , \tag{3.6}$$

where I_ϕ is the magnitude of I_n at angle ϕ in *fig 3-5c*.

When passing from one medium to another having a different refractive index, as in going from air to water, light partially reflects and partially penetrates at an angle that depends upon the refractive indices *(eq 3.3)* of the two media, the change in direction being called *re-*

Fig 3-5. Types of reflectance. On a micro-smooth surface (a, left), the angle of re-flection e-quals the an-gle of inci-dence on a gross level, but on a mi-cro-rough sur-face (b, right) the equality of angles causes multiple small reflections re-sulting in gross diffu-sion of the re-flected light. Ideal diffusion (c, bottom) follows Lam-bert's law (eq 3.6).

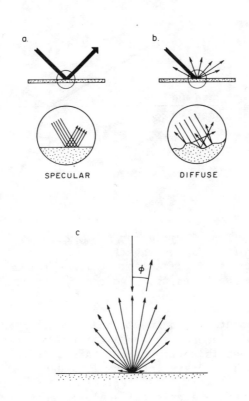

fraction. Wavefront theory accurately predicts that

$$n_2/n_1 = \sin\theta_1/\sin\theta_2 , \qquad\qquad (3.7)$$

where θ_1 is the angle of incidence, θ_2 the angle of re-fraction and n the index of refraction, as shown in *fig 3-6* (p. 66). Because $n_{air} \doteq 1$, the ratio of sines in *eq (3.7)* provides the refractive index of denser materials such as water ($n_{water} \doteq 1.3$, depending upon temperature and dissolved materials, *i.e.*, depending upon density). Refraction is important to fish in detecting predatory seabirds in the air above the surface, and is also the basis of lenses, such as those in light-emitting organs (*ch 4*) and eyes (*ch 5*).

A peculiar thing happens when light passes from one medium into a less dense one, as in passing from water in-to air. The refracted light is bent *away* from the normal,

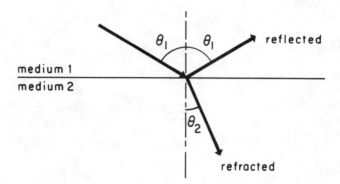

Fig 3-6. Refraction of light as it leaves a less dense medium (1) and passes into a more dense medium (2). The angles of reflection (θ_1) and refraction (θ_2) are related by the refractive indices of the media by eq (3.7).

and at a certain *critical angle* of incidence numerical solutions of *eq (3.7)* for the angle of refraction yield impossible values. At this critical angle, no refracted light emerges into the less dense medium; instead, there is *total internal reflection* within the denser medium. For water, the critical angle is about 49°, and for the glass-air interface it is about 42°--a fact utilized in the design of 45-90-45° prisms used as perfect mirrors to change the direction of light. Because of reflection at the air-water interface, fish face a potential problem of confusing real images of objects above or below the surface with reflected images of objects in the water. Moody (1975) has shown that rosy barbs, however, can make the discrimination with ease.

The reflected light *(fig 3-6)* is partially polarized in the plane parallel to the surface between the media, and when $\theta_1 + \theta_2 = 90^{\circ}$, the reflected light is completely polarized. Some materials also polarize light by transmitting vectors in one plane but absorbing those in others. Since environmental glare from water and wet roads is horizontally polarized reflectance, sunglasses that transmit only vertically polarized light cut glare without reducing other light substantially.

Most transparent substances such as glass refract light differentially throughout the visible spectrum, such that

$$n \uparrow \nu . \qquad (3.8)$$

Newton capitalized on this fact when he used a glass prism to spread white light into its component frequencies (*re-fractive dispersion*). However, this same property is an important hindrance in the manufacture of lenses because it causes *chromatic aberration*: focusing of different frequencies to different places in space. Chromatic aberration of the eye produces a depth-illusion based on color *(ch 5)*.

scattering and polarization

Equation (3.4) is not a complete statement of the fate of incident light because some light may be scattered within the substance through which it passes. Scattering may be thought of as multiple reflection by small particles. When the particles are very small ($\leq \lambda/10$) they scatter light nearly independently of one another according to

$$s_{Ray} \propto (n-1)^2/\lambda^4 , \qquad (3.9)$$

where s_{Ray} is the coefficient of *Rayleigh scattering*. *Equation (3.9)* means that short wavelengths (high frequencies) from the "blue" end of the visible spectrum will be scattered strongly, whereas long wavelengths (low frequencies) from the "red" end will penetrate strongly. For this reason, we see penetrating light from the rising sun as very orange and the back-scattered light to the west as deep blue. Scattering is not precisely proportional to the inverse of the fourth power of the wavelength *(eq 3.9)* —as sometimes stated—because the n is spectrally dependent *(e.g., eq 3.7)*. Electric vibrations in the particles polarize light scattered normally to incident radiation, as may be appreciated by looking at the sky $90°$ from the sun and rotating Polaroid sunglasses.

When scattering particles are large ($> 25\lambda$) ordinary geometrical optics applies, but when the range is between this and Rayleigh scattering the effect is *Mie scattering*, which has complicated mathematical properties. Mie scattering is due to the interaction among light waves that have been reflected, refracted and otherwise affected.

At small particle sizes (relative to the wavelength of
light), Mie scattering occurs in all directions relative
to the incident light-path, although the scattering is
never uniform as in Rayleigh scattering. As the particle
size increases, Mie scattering is directed forward to an
increasing degree until nearly all scattered light is
directed forward when the radius of the particle is about
the size of the wavelength. The Mie scattering coefficient may be summarized as

$$s_{Mie} \propto \gamma^2 M(n, \gamma/\lambda) , \qquad (3.10)$$

where γ is the radius of the scattering particle and M is
a dimensionless oscillatory function. If the particle
size and index of refraction are held constant, the effect
of wavelength on Mie scattering may be appreciated. At
very short wavelengths (large γ/λ values), M has a limit-
ing dimensionless value of about 2. As the wavelength becomes longer, M rises to higher values and falls to about
2 in oscillations that increase until a final peak near
$\lambda = \gamma$. After this final peak, Mie scattering falls monotonically to zero, as shown in *fig 3-7* for the refractive

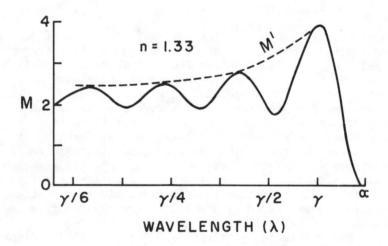

*Fig 3-7. Wavelength-dependency of Mie scattering depends
upon the function M (eq 3.10). Wavelengths are expressed
in terms of the radius (γ) of the scattering spherical particles, which have the refractive index (n) of water in
this exemplary function.*

index of water. The refractive index changes the oscillatory function only quantitatively, and a given maximum
value of M occurs at increasingly longer wavelengths as n
increases.

The spectral effects of Mie scattering are very complex and depend on the size of the particle in relation
to the wavelength of light, as may be appreciated with
some quantitative examples in reference to fig 3-7. Sup-
pose $\gamma = 400$ nm; then $s_{Mie} \uparrow 1/\lambda$ (for $400 \leq \lambda \leq 750$ nm),
and the effect is somewhat similar to that of Rayleigh
scattering $(eq$ 3.9), in that scattering decreases monotonically with wavelength. However, in Mie scattering,
light of wavelength near the particle-size diameter is
scattered forward, and hence does not cause the strong
filtering-like effect of Rayleigh scattering.

Continuing with examples, at $\gamma = 550$ nm, $\gamma/2 = 225$ nm
so that by reference to fig 3-7 Mie scattering increases
with wavelength from 400 to 550 nm in the visible, then
decreases from 550 to 750 nm. Or, with $\gamma = 750$ nm, $\gamma/2 =$
375 nm and the entire visible spectrum is contained between $\gamma/2$ and γ, meaning that Mie scattering increases
monotonically with wavelength. At the upper limits of
particle size causing Mie scattering, $\gamma \doteq 18$ μm, so that
$\gamma/6 \doteq 3$ μm and the entire visible spectrum is off the
graph to the left of fig 3-7; Mie scattering is therefore
almost constant with wavelength. Finally, as a last example, consider $\gamma = 900$ nm. Then the visible spectrum
lies between $\gamma/2.25$ and $\gamma/1.2$, meaning that Mie scattering
from short to long wavelengths in the visible first decreases to a minimum, then increases again.

Therefore, even though the general trend of peaks in
fig 3-7 is to increase with wavelength (the broken line
labelled M'), the spectral effects of Mie scattering within the visible spectrum can be extremely varied. Furthermore, most scattering aerosols contain a variety of particle sizes, often of mixed materials with differing refractive indices, so that virtually any kind of spectral
effects are possible. This means that effects of air pol-
lution in the atmosphere (mentioned again later in this
chapter), and transmission of light between sender and receiver in a scattering medium such as fog, must be investigated empirically in given cases.

diffraction and interference

Light spreads out in space after passing through a small hole, and at the edge of every obstacle the light-path is bent. Such _diffraction_ can be seen by looking at a distant streetlamp through a window screen: light is bent at the four edges of the square holes of the screen, and hence spreads out vertically and horizontally to form a cross.

Light rays from the same source that reach the same locus in space by different routes have waves that may be in or out of phase. These waves interact, and if they are in phase they add (_constructive interference_), whereas if they are out of phase they cancel (_destructive interference_). The result of diffraction from two slits or pinholes is patterned light due to such _diffractive inter-ference (fig 3-8)_. Rayleigh was the first to realize that

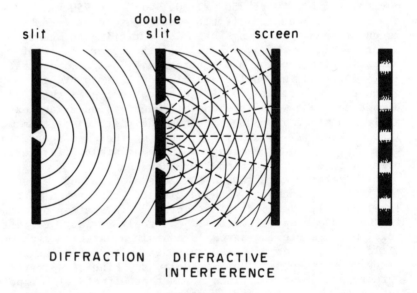

slit   double
                         slit   screen

DIFFRACTION   DIFFRACTIVE
                   INTERFERENCE

Fig 3-8. Diffraction and interference. Light from a single source spreads out after passing through a slit (diffraction). The wavefronts propagated from two slits interfere, reinforcing along some loci (dashed lines) and partically cancelling elsewhere. At right is the banded pattern of light and dark as seen on the screen.

diffraction and subsequent interference of light on the
shadowed side of an object should create a light spot
in the shadow. He demonstrated this counter-intuitive
principle with a penny placed in a shaft of sunlight, although the phenomenon is a subtle one and difficult to
see convincingly without a controlled laboratory setup.
Rayleigh's penny is analogous to the opaque portion between the double slits in *fig 3-8*: there is a bright spot
centered behind the portion. The slits in *fig 3-8* could
also be spacings between small rods aligned in a row,
causing interference of light waves analogous to the way
in which rows of pilings cause interference of surface
waves on water. Light waves from different sources do not
ordinarily show interference because of constantly changing
phase relationships.

Interference also occurs without diffraction when two
partially reflecting surfaces are arranged parallel (*thin-
layer interference*), as in the inner and outer surfaces of
a soapbubble. Some of the light that penetrates the first
surface is reflected back by the second. If the spacing
between surfaces is exactly one-half the wavelength, the
interference of incoming and reflected light is completely destructive (*fig 3-9*). Destructive interference occurs

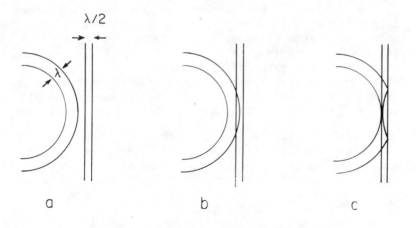

*Fig 3-9. Interference in thin films such as soapbubbles.
The parallel vertical lines represent the inner and outer
surfaces of the film and two wavefronts are shown approach-
ing from the left. The film is one-half* (cont. p. 72)

at a series of spacings ($\lambda/2$, $3\lambda/2$, $5\lambda/2$, *etc.*). Thin-
layer interference is the basis of high reflectivity in
fish scales (see *ch 4*, below).

Diffractive and thin-layer interference are both used
as principles for creating dispersion by means other than
a prism (refractive dispersion). A plate ruled in precisely spaced lines (*diffraction grating*) or two metal
films spaced by increasing distance along one dimension
(*interference wedge*) are both used in monochromators:
devices that disperse white light and isolate the component wavelengths for experimental use.

iridescence and dichromatism

The foregoing optical principles explain the variation of surface colors under different viewing conditions.
Some surfaces appear multicolored, or appear differently
colored at different viewing angles. These phenomena are
almost always due to interference (*e.g.*, *figs 3-8* and *3-9*)
and the resultant coloration is *iridescence*. Repetitive
elements of microstructure of a substance give rise, often
in complex geometries, to diffractive iridescence, as in
the feathers of some birds (*ch 4*), whereas spaced reflecting layers give rise to thin-layer iridescence, as in the
scales of fishes. The latter structure shows different
colors either (1) because the spacing varies, reinforcing
some wavelengths here and others there, or (2) because
light strikes the substance at different angles and hence
has different lengths of light-path between the layers, so
that one wavelength is reinforced at one angle and another
wavelength at another angle. The iridescent sheen of
mother-of-pearl in molluscs is probably due to both phe-

wavelength in thickness. The first wavefront (a) to reach
the film is partially reflected from the first surface (not
shown) and partially penetrates to strike the second surface (b), where it is partially reflected back. By the
time the second wavefront reaches the first surface (c),
the reflected first wavefront reaches the same point and
the second cancels the first moving in the leftward direction. Light of wavelength λ is therefore eliminated
from the reflection in the leftward direction. However,
the first wavefront is re-reflected toward the right where
it combines with the second wavefront constructively to e-
merge in the rightward direction. Thin films therefore
selectively remove wavelengths from reflection.

nomena acting simultaneously.

Perhaps commonly occurring but not commonly noted are substances that appear two different colors, or vary continuously between two colors. Such coloration is due to a variety of causes variously called *dichromatism* or dichroism. The molecular structure of some materials differentially absorbs photons of different vector-orientations; therefore, the materials transmit polarized light of different colors depending upon the plane of polarization. Any substance that polarizes light by selective absorption of vectors is often referred to as *dichroic*, even if the absorption is spectrally neutral so that the substance appears the same color (but of different intensities) when irradiated with light of different planes of polarization. Because of this usage of dichroism, the generic term dichromatism is preferred for materials that appear in two colors.

Dichromatism may be due to causes other than dichroism. Commonly a substance will have two spectral bands absorption. Suppose a given thickness (or concentration of solution) absorbs 30% of the light ($a = 0.3$, *eq 3.4a*) in one spectral band and 65% in the other, so that the transmissions are $t_1 = 0.7$ and $t_2 = 0.35$. If the thickness is doubled, these values become $t_1 = 0.7^2 \doteq 0.5$ and $t_2 = 0.35^2 \doteq 0.12$. Therefore, in the first case of single thickness, $t_1/t_2 = 2$, whereas doubling the thickness yields $t_1/t_2 = 4$. That is, the relative contribution of the two spectral bands to the transmitted light varies with the thickness or concentration, hence so does the apparent color. One expects such *absorptive dichromatism* to occur with substances that appear green because green is the color sensation evoked by energy in the middle of the visible spectrum; therefore, green is absorptively created by absorption at spectral extremes (two absorption bands).

Absorptive dichromatism also occurs with reflected light viewed from different angles. This possibility seems counter-intuitive because we think of reflection as a simple bouncing of light from a surface (as in *figs 3-5* and *3-6*). Actually, some of the reflected light is absorbed and immediately re-radiated by molecules and other of it may be scattered back within a substance. Therefore, incident light penetrates substances and is partially absorbed before being propagated back as gross reflection. Whenever incident light penetrates materials to different degrees, absorptive dichromatism may occur. For example, if light penetrates on the average to a particular micro-

depth beneath the surface, it has a shorter path within
the surface when incident at high angles than low ones.
Similarly, absorptive dichromatism is responsible for
differences between the reflectance and transmittance
spectra of some substances.

Environmental Light

The foregoing principles of the interactions of light
with matter lay foundations for discussing the promulga-
tion of optical signals *(ch 4)* and their reception *(ch 5)*,
but are also of immediate interest in describing more
closely the real optical environment for signaling. Bioluminescent organisms excepted, almost all light used in
animal communication originates with the sun, which must
therefore be the starting place for discussing environmental light. Sunlight is altered in the atmosphere, and
then again in terrestrial and aquatic habitats, where the
media of transmission differ importantly in properties
relevant to the design of optical signals.

irradiance on the earth

The spectrum of sunlight measured from a rocket high
above the earth's surface resembles that of an ideal
black-body radiator *(fig 3-10a)*. The peak irradiance-
density is near 0.5 µm (500 nm), which is near the peak
sensitivity of the human eye *(ch 5)*, but sunlight also
contains appreciable ultraviolet radiation (λ < 400 nm)
and infrared radiation (λ > 750 nm). Many animals (particularly arthropods) are known to see into the near-UV,
but only pitvipers and a few other animals can sense the
near-IR *(ch 5)*.

The irradiance at the surface of the earth differs
from pure sunlight because of atmospheric effects *(fig
3-10b)*. The spectrum shows bands of low irradiance in
the IR due to absorption by water vapor in the atmosphere.
Rayleigh scattering shifts sunlight toward longer wavelengths, but also causes the entire sky to become luminous
with a pronounced bluish coloration. (Space as seen from
the moon is black because the moon has no atmosphere.) An
irradiance spectrum measured just above the surface of the
earth on a large, flat plane results from the combination
of direct solar and sky-scattered radiation. When the sun

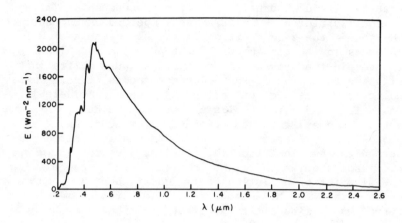

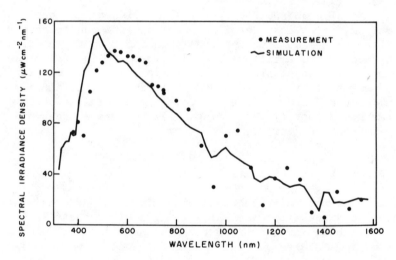

Fig 3-10. Irradiance-density of sunlight measured from a rocket (above) and at the earth's surface (below). (a) The upper curve is a proposed standard average (Thekaekara, 1972). (b) The lower curves are measurements made by the author in Panama (dots) and the spectrum simulated (line segments) by the model of McCullough and Porter (1971) for the same location and time (Barro Colorado Island, April 1973, noon, clear day).

is low in the sky, around the time of sunset and sunrise, its light has a longer path through the atmosphere so that

scattering and absorption are more pronounced. Similar
seasonal differences in irradiance spectra also occur with
the sun at the same position in the sky because of differences in the atmospheric light-path *(ch 7)*. Finally,
changes in the atmospheric conditions cause shifts in the
light reaching the earth. Clouds composed of large water
molecules cause Mie scattering and absorb considerable
radiation, and more complex effects may be due to air pol-
lution such as from industrial smoke or eruption of vol-
canoes. A fire raging in Albert, Canada in September 1950
turned the sky in Edinburgh, Scotland deep indigo, where
R. Wilson studied the light in great detail, calculating
the diameter of polluting particles at about 1 µm based
on equations for Mie scattering. Similar phenomena occurred world-wide after the great eruption of Krakatoa
in 1883 (Ruechardt, 1958).

McCullough and Porter (1971) published a computer al-
gorithm, based on a detailed model of atmospheric conditions, that simulates spectral irradiance on the earth's
surface. This ambitious tool delivers simulations at any
geographic coordinates, any elevation, any time of year
and any time of day, with added provisions for relative
humidity and amount of cloud cover. The simulation is
remarkably good *(fig 3-10b*, p. 75), and has subsequently
been improved (Porter, *pers. comm.*).

shadows on the substrate

The sun's disk subtends an angle of about a half-de-
gree as viewed from the earth (31' 28" in January when
the earth is farthest from the sun and 32' 30" in June
when it is nearest). This subtended angle expressed in
radians averages about $\Theta = 1/107.5$ rad. Any circular object smaller than the sun but subtending the same angle
exactly blocks the sun's image, and hence casts a con-
verging umbral shadow *(fig 3-11)*. It follows from the
definition of a radian that an object of diameter δ_0 casts
an umbra that converges at a distance about

$$D = \delta_0/\Theta \, , \qquad\qquad (3.11)$$

because the diameter is very nearly an arc of a circle
with radius D, as shown in *fig 3-12a*. Therefore, the
length of the umbra is about 108 diameters of the object,

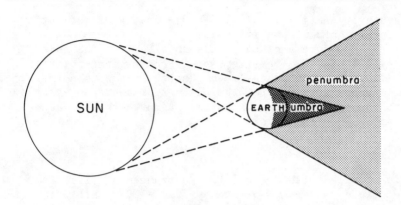

Fig 3-11.   Types of shadows cast by an object in sunlight
(not to scale).

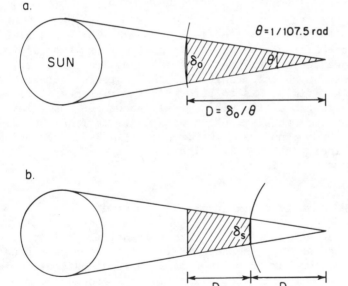

Fig 3-12.   Umbral shadows of objects.   The length of an
umbra (a) is about 107.5 times the diameter (δ_0) of the
object casting it, and the diameter of the shadow (δ_s)
cast by an object closer to the substrate than $107.5\delta_0$
is linearly proportional to the distance (D_s).

and every animal within 100 body-lengths of the earth
casts a shadow upon its surface.

Introduce a plane at distance D_s from the object
and distance D_c from its umbra's convergence-point (*fig
3-12b*, previous page), such that the plane is normal to
the umbra's axis. Let the plane represent the surface
of the earth or other substrate upon which the shadow is
cast. From the figure,

$$D_c = \delta_s/\Theta \; ,$$

in parallel with *eq (3.11)*. Rearrangement yields $\delta_s =
D_c\Theta$, and since $D_c = D - D_s$ by *fig 3-12*, substitution gives

$$\delta_s = D\Theta - D_s\Theta \; .$$

Finally, by *eq (3.11)* $\delta_O = D\Theta$, so that

$$\delta_s = \delta_O - D_s\Theta \; . \tag{3.12}$$

This little exercise in algebra secures the conclusion
that the size of the shadow is a linear function of the
distance to the substrate. The shadow approaches the
size of the object itself when the distance decreases to
zero, and diminishes to zero when the distance increases
to the length of the umbra (*i.e.*, about 108 body-diameters).
And, as the umbra decreases with distance from the substrate, the penumbra grows in size (*fig 3-11*), creating a
hazy, indistinct shadow. The importance of shadows cast
by animals is discussed in *ch 6*.

ambient irradiance in terrestrial habitats

Even a desert animal is not irradiated solely by sunlight and skylight because it is not a plane on the earth's
surface. Rather, an animal is a three-dimensional object
that receives reflected light from the ground and other
objects in the environment. In vegetated habitats these
objects are in part plants, which not only reflect light
but also filter light through the leaves. *Figure 3-13*
shows the irradiance spectrum in a forest in Panama, with
minima near 425 and 675 nm due primarily to absorption by
chlorophyll. Although there is considerable transmission
of long wavelengths, the eye is so insensitive to the far-

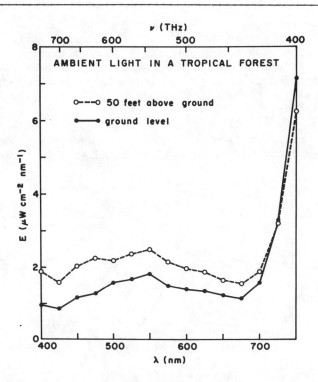

*Fig 3-13. Spectral irradiance-density in a forest, meas-
ured by the author in Panama at two different heights
(Barrow Colorado Island, 18 April 1973). The maximum near
550 nm is due primarily to absorption by chlorophyll at
shorter and longer wavelengths, giving ambient light a
greenish hue. The high irradiance at very long wavelengths
occurs where the human eye is almost insensitive to light.
Irradiance-density is stronger high in the forest, but its
spectral distribution is similar to that at ground level.*

red part of the spectrum (*ch 5*) that forest light has a
distinctively greenish hue due to the maximum irradiance
near 550 nm. Moving upward within the forest increases
the ambient light but does not appreciably alter its
spectrum.

The ambient irradiance in terrestrial habitats is so
complexly determined and so poorly studied that it cannot
be modeled accurately at present and so must be determined
empirically. Burtt (1977) made measurements in a variety
of habitats in Minnesota and found that the primary spec-

tral differences were between broadleaved and coniferous
forests. Light in coniferous forests is governed more by
reflection from opaque materials such as branches and
needle-leaves than by transmission through broad leaves
as in hardwood forests. Consequently, the coniferous
spectra tend to be flatter than those in *fig 3-11*, and
the ambient light is more neutral than greenish. The
absolute levels of irradiance, however, vary with the
height and density of the trees. *Figure 3-14* summarizes
sources contributing to ambient irradiance in terrestrial
habitats.

*Fig 3-14. Principal sources of irradiance on an animal
in a terrestrial habitat. Direct sunlight (I_o) or sunlight scattered in the sky (I_s) may fall on the animal
directly, be filtered by leaves and other objects (I_t)
or reflected (I_r) from plants or other surfaces such as
the ground. Diagramed is the male painted bunting, one
of the most colorful birds in North America, having a
blue head, red breast and rump, yellowish back and green
wing-patches.*

ambient irradiance in open ocean

The ambient irradiance in aquatic habitats is even more complexly determined than that of terrestrial habitats because of added factors. The irradiance falling on the surface of a woodland stream, for example, is subject to all the factors shown in *fig 3-13*, and then to additional factors within the new medium of the water itself.

It is useful to begin with the open tropical ocean where direct sunlight and skylight fall upon the surface of relatively clear water. Many factors in the irradiance of oceans are counter-intuitive so that it is useful to consider factors one by one. The ocean appears blue from above, and popular articles sometime ascribe this coloration to backlight from Rayleigh scattering. If that were true, long wavelengths should penetrate the ocean in the way that long wavelengths from the rising and setting sun penetrate the atmosphere. However, SCUBA divers know that long wavelengths disappear *first* in ocean depths: light becomes more bluish rather than reddish beneath the surface. Furthermore, inspection of spectral irradiance curves suggests that the bluish light above and beneath the surface of oceans does not have the expected Rayleigh scattering characteristics.

Spectral absorption measurements on water give conflicting results because of materials dissolved in it, but it does appear that absorption rises with increased wavelength:

$$\alpha_{water} \uparrow \lambda \ . \tag{3.13}$$

Therefore, light in the ocean tends to lose energy from the long wavelengths ("red" end of the visible spectrum) toward the short wavelengths. The blue seen from above the surface is therefore probably due to backlight of Mie scattering or gross scattering, or both, and not primarily to Rayleigh scattering.

Until recently, available data have been spotty, but the study by McFarland and Munz (1975a) provides support for the foregoing reasoning as well as a useful bibliography to older data. *Figure 3-15a* shows their sample results of the spectra of irradiance falling upon a tropical sea, expressed in quantal rather than energy terms (see *eq 3.2*, p. 56). The observed quantal spectra are relatively flat on both clear and overcast conditions, but 3 m below the surface the irradiance from above shows a

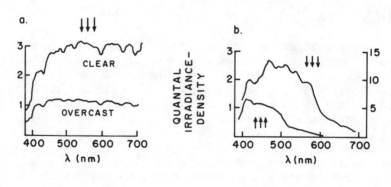

ABOVE SURFACE   3M BELOW SURFACE

Fig 3-15. Spectral irradiance-density in an ocean (after McFarland and Munz, 1975a). At left (a) is the downward quantal irradiance-density above the ocean under clear and overcast conditions, and at right (b) is the downward irradiance (top curve, left axis) and upward irradiance (bottom curve, right axis) at a point 3 m below the water's surface. All vertical axes are in units of quanta cm^{-2} s^{-1} nm^{-1} and must be multiplied by these factors for absolute values: 10^{14} (left axes of both diagrams) and 10^{12} (right axis of right diagram).

dramatically shifted spectrum *(fig 3-15b, top curve)*. The bluish underwater light shows a similar spectrum when the detecting instrument is pointed downward *(fig 3-15b, bottom curve)*, due to the back-scattering of the penetrating light that is not absorbed by the water.

McFarland and Muntz *(op. cit.)* went on to measure the quantal irradiance spectra at many different directions from the same place beneath the surface. As expected, irradiance from the sun's direction is stronger and hence its long wavelengths are more pronounced, but for more than a hemisphere around a point beneath the surface the irradiance spectra are highly similar to the lower curve in *fig 3-15b*. As one descends in the water-column two predictable effects take place: (1) the ambient spectrum becomes increasingly narrow, shifting toward short wavelengths (bluish light), and (2) the ambient spectrum becomes similar in all directions. McFarland and Muntz state that the minimum depth at which the directional similarity takes place varies with water conditions, and may be deeper than 100 m in most clear tropical seas. Tyler (1960) made a detailed study of the total irradiance in various directions from loci at increasing depths in an inland lake, finding results generally similar to those of McFarland and Munz for the oceanic environment.

ambient irradiance in complex aquatic habitats

Not all aquatic animals conduct pelagic lives in open oceans. Important ecosystems associated with coral reefs and shallow seas contain many of the fishes and other animals that use optical communication extensively. McFarland and Munz (1975a) found that in shallow waters the reflection of light from the bottom substrate played an important role in irradiance from below an animal. As expected, the shallower the water, the more intense the reflected contribution to irradiance, with corresponding shifts in spectrum away from short-wavelength-dominated light *(fig 3-16a)*. Furthermore, the type of substrate

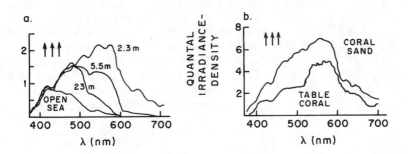

1 M BELOW SURFACE   1 M BELOW SURFACE

Fig 3-16. Upward irradiance in shallow sea (after McFarland and Munz, 1975a). At left (a), the curve broadens as the bottom becomes closer to the surface, due to reflection from the bottom. At right (b), the reflectivity of the bottom surface influences the spectral distribution of upward irradiance. Axes are in units of quanta cm^{-2} s^{-1} nm^{-1} and must be multiplied by correction factors for absolute irradiance-density: 10^{13} (left diagram and top curve of right diagram) and 0.33×10^{13} (bottom curve of right diagram).

affected the spectral distribution and absolute levels of
reflected light (*fig 3-16b*, previous page). One is left,
therefore, with the same situation resulting from considerations of ambient irradiance in terrestrial habitats:
the factors are too many and complex to model accurately,
and empirical measurements must be made for given cases.

In estuarine and inland waters, the ambient light is
further complicated by *turbidity*: absorption and scattering due to suspended materials such as mud particles. The
spectral effects of turbidity probably vary greatly with
the nature of the dissolved materials, but Luria and Kinney (1970), Jerlov (1968), Hutchinson (1957) and others
find that

$$a_{\text{turbidity}} \stackrel{\uparrow}{=} \nu , \qquad\qquad (3.14)$$

which is disconcertingly the opposite effect of absorption
by water alone (*eq 3.13*, p. 81). (In *eq 3.14*, *a* may be
due to scattering as well as true absorption, and Mie
scattering of *eq 3.10* could play an important role.) Furthermore, turbidity due to aquatic organisms such as phy-
toplankton may have quite different spectral effects on
ambient irradiance, absorption by chlorophyll creating a
greenish hue similar to that found in forests.

Finally, in inland waters, the important effects of
vegetation and other factors of surrounding terrestrial
habitats affect the spectral quality of light before it
strikes the water. It seems a conservative conclusion to
state that ambient irradiance in aquatic habitats (outside open ocean) is quite complexly determined, studies
to date having yielded few trusty generalizations.

Transmission

Most optical signals of animals consist of ambient
light reflected from the sender's body or another object
controlled by the sender. *Chapter 4* concerns details of
the reflection processes, and this section is devoted to
generalities of how the reflected signal is affected during transmission from sender to receiver.

signal radiance and irradiance

An optical signal is created when ambient light (I_0)

---

irradiates a surface of given spectral reflectivity (r)
and is reflected as signal-radiance $(I_r$ or simply $I)$ according to *eq (3.4a)*. Most animal surfaces used as sig-
nal-surfaces have diffuse reflectance, and hence approximate Lambert's law *(eq 3.6)*. Therefore, the radiance of
a signal at a given angle of view is

$$I \propto rI_0 \cos \phi \ . \qquad\qquad (3.15)$$

The signal surface acts as a secondary source radiating
light that falls on another surface (*e.g.*, a receiver's
eye) as signal-irradiance (I'). The relative area of the
signal-surface in the receiver's field of view decreases
by the square of the distance from the surface. The surface continues to appear just as bright per unit area (in
a perfectly transparent medium), but since its relative
area decreases, the total irradiance arriving at the viewer decreases. Put simply, signal-radiance (I) remains
constant and signal-irradiance (I') decreases with distance.

   Only for infinitely small surfaces (point-sources)
will signal-irradiance follow the inverse-square law:
$I' = I/d^2$, where d is the distance between source and irradiated surface. Irradiance from an infinite line source,
approximated by a very long fluorescent tube viewed from
afar, decreases linearly with the distance $(I' = I/d)$, and
irradiance from an infinite plane is independent of distance $(I' = I)$. Finite radiant surfaces therefore fall
somewhere between infinite planes and point-sources, approximately the former when they are large and the latter
when they are small, so that one may state only that

$$I' \uparrow I/d \ , \qquad\qquad (3.16)$$

which is none-the-less a possibly non-obvious phenomenon
in a clear medium.

distortion and noise

   The transmission of the signal from sender to receiver encounters two kinds of distrubances in the channel: distrubances by the medium itself (air or water) and
disturbances by objects in the medium, primarily opaque
objects such as vegetation, coral heads and so on. All

of the factors that influence the spectral quality of environmental light enumerated above can affect the light
transmission from sender to receiver. In actuality, the
communicative distance is usually so small that only some
of the factors are really important. Air is ordinarily
transparent, but fog can alter the spectral quality of
light by Mie scattering (eq 3.10) and alter the image of
the signal by diffuse Mie and gross scattering, thus seriously disrupting signals. Turbidity in water (eq 3.14)
similarly disrupts the image-clarity and spectral distribution of light reaching the receiver. Whether these disruptions constitute mere distortion or important noise
(see ch 2) becomes a consideration of the optical design
of signals (ch 7). The point here is that many of the
same factors that influence ambient light on the signal
surface, discussed in foregoing sections, also influence
the transmission of light from sender to receiver.

Disruption of the signal by opaque objects in the
medium is a serious problem for communication, and one
that has received little systematic attention. As an example of the genre of unstudied problems that exist, consider whether or not a receiver can see an entire signal-
object, such as a colored patch on an animal or a gesture
with a appendage. For simplicity, consider the object to
be circular with area A_o. Then its apparent area is A_o/d_o^2,
where d_o is the distance from object to viewer. If an environment such as a forest has circular "holes" in the vegetation mass through which one can see, the apparent area
of a hole will be A_h/d_h^2, where A_h is the actual area and
d_h the distance from the observer. Only when

$$A_h/d_h^2 \geq A_o/d_o^2 \qquad\qquad (3.17)$$

will the observer see the entire signal-object. Since A_h
and A_o are constant (for a given hole and object), eq (3.17)
is encouraged toward the favorable inequality by minimizing
d_h and maximizing d_o. Systematic investigation of such
problems in the transmission of optical signals in real
environments might greatly aid the understanding of the
size and shape of optical signals (see ch 7).

A further problem arises in the transmission of signals in that they must be discriminable from other visual
arrays. Therefore, signal-objects should be of different
size, shape, color, etc., from other objects in the environment, which constitute an important source of noise.

It seems likely, for example, that signals of terrestrial
animals tend not to be green simply because there is so
much green background against which the signals are seen
by the receiver. The effects of these kinds of noise problems on the design of optical signals is the subject of
ch 7.

Overview

In order to transmit information, optical signals
must have variation. The relative intensities of different spectral frequencies in various spatial arrays provide a nearly boundless number of possible signals, which
may be rapidly modulated, so that the informational capacity of the optical channel *per se* is effectively boundless. Limits, if they exist, must lie with senders and
receivers, and hence it is necessary to review the physical phenomena having to do with the interaction of light
with matter: refractive index, radiance and irradiance,
monochromaticity and spectral complexity, polarization,
types of emission, absorption spectra, reflectance and
transmittance, specular and diffuse reflectance, refraction and dispersion, types of scattering, shadows, diffraction and interference, iridescence and dichromatism,
turbidity and geometric considerations of radiation. Because most optical signals of animals utilize reflected
sunlight, it becomes relevant to know how ambient light
is determined in different habitats. Differences in ambient light of terrestrial habitats are expected to correlate with time of day, time of year, geographic location,
altitude, weather (*e.g.*, fog), open areas *vs* woods, and
within the latter, coniferous *vs* broadleaved woods, as
well as height above ground. Ambient light in most aquatic habitats is possibly even more complexly determined,
being affected by time of day, time of year, geographic
location, weather, depth beneath the surface, water-depth
itself, turbidity and so on. Once ambient light has been
reflected from a signal-surface only a few of the foregoing factors, such as fog and turbidity, affect its transmission to the receiver. However, image-clarity becomes
important, so opaque objects in the channel (*e.g.*, plants)
undoubtedly constitute noise. Finally, the similarity of
signals with other optical arrays presents possible problems in the transfer of information from sender to receiver.

Recommended Reading and Reference

Most of the material in this chapter concerns straightforward physical optics that may be found in most college textbooks on physics; the reader without calculus at his command should choose a text that explains physics with algebra only. The entire September 1968 issue of *Scientific American* is devoted to light and its interactions with physical and biological matter. Clayton (1970) is the first of two small volumes of introduction to *Light and Living Matter*, presenting the physical basis of absorption and emission processes. An old, but useful and inexpensive reference to have at hand in the 1963 Dover (revised) edition of Monk's *Light*. A delightful book about natural phenomena of environmental light is Minnaert's (1954) *Nature of Light and Colour in the Open Air*. Many other volumes on physical optics are available. A treatise on scattering is provided by van de Hulst (1957), and there are good discussions of this and other topics in Robinson's (1966) *Solar Radiation*; Coulson (1975) includes much new data.

Chapter 4

THE SENDER

*Every true or inherited movement of expres-
sion seems to have had some natural and independent origin. But once acquired, such
movements may be voluntarily and consciously employed as a means of communication.*
                                        —Darwin

The sender transmits information by creating variety
in light that reaches the receiver *(ch 2)*. The sender's
abilities to create variety therefore constrain the design of optical signals and their capacity to carry information. This chapter concerns three principal aspects
of the sender's abilities: basic mechanisms for producing
and reflecting light, a classification of gross aspects
of signals based on those mechanisms (taken from Hailman,
1977a) and some evolutionary constraints on optical signals.

Mechanisms of Signal-generation

Animals generate optical signals by creating variation in the temporal and spatial patterning of light,
either by producing light themselves or by reflecting
ambient light. Temporal modulation of reflected light
depends ultimately on animal movement and when the light
is reflected from the animal's own surface, on reflective
characteristics of the integument. This section therefore briefly reviews the principal aspects of bioluminescence, animal movement and the bases of animal reflectivity.

bioluminescence

Animals generate their own light in one of three ways. Some have *extracellular organs* that secrete luminescent materials into the sea around them, as is known from a shrimp and from a squid that squirts a glowing cloud as part of its escape mechanism. Other animals secrete a *luminescent slime* over their bodies, and this process involves more than simple dumping of chemicals into the water, although it has not been well studied. The third, and usual, mechanism is *intracellular* bioluminescence, in which light-producing cells are collected into a functional organ.

Photic organs are known from many of the major groups utilizing optical communication, including cephalopods, crustaceans, insects and fish. Such organs differ among species, but are usually much more than a collection of bioluminescent cells. The organs may have light-absorbing layers that can shield the luminescent structure, light-reflecting layers to enhance propagation of light away from the animal, filtering tissues to change the color of luminescence and refractive bodies resembling lenses, in addition to nerve supplies and in some cases movable covering structures. In fact, many such photic organs have the superficial appearance of an eye. The ultimate color of observed luminescence is determined by the molecular mechanism plus the reflecting layer and filters.

Not all light-emitting animals are actually bioluminescent themselves because some contain luminescent bacteria--the only case known to me where one symbiont of a pair derives purely communicational advantage from the relationship. It is often difficult to tell whether the animal or its symbionts are generating light without detailed anatomical investigations. Although most luminescent bacteria glow chronically, certain fishes can modulate the emitted light by moving a cover over their microorganisms.

The chemical mechanisms of bioluminescence are incompletely understood and undoubtedly vary somewhat among species. A pigment-like, heat-stable molecule of low molecular weight known generically as *luciferin* emits light when it is oxidized. The oxidation is mediated by a large, heat-labile, water-soluble catalyst, chemically identified in some cases as an enzymatic protein and hence known as *luciferase*. The reaction always requires water and oxygen, although the partial pressure of oxygen needed may be

quite low, and in some cases other chemical constituents
may be needed. Bioluminescence has a longer latent period than the radiation-induced processes of some fluorescence and phosphorescence *(ch 3)*, and emits light over a
longer time-span once activated.

The spectrum of bioluminescence usually has a band-
width of between 80 and 200 nm, and always occurs in the
visible spectrum from about 415 to 670 nm, often centered
about 480-510 nm. A single animal may bioluminesce in two
different colors; in some cases this is due to a single
mechanism rather than two, depending upon elaborate anatomical specializations associated with the basic light-
producing mechanism.

The temporal control of light-emission, which is the
primary information-carrying variable in bioluminescence,
has various sources. In some cases, the neurally controlled
production or release of luciferin and luciferase determines when luminescence shall take place. In other cases,
the necessary water or oxygen is controlled, and in some
cases the metabolic machinery altering paths producing
precursors or other necessary chemicals is the variable
under physiological control. In addition, animals that
possess an opaque covering for a bioluminescent structure
are able to open and close that covering by conventional
muscle contraction. All these factors lead one to expect
good temporal control over light-emission, which seems to
be the case in at least some well-studied species such as
fireflies *(e.g., Lloyd, 1977)*.

muscular movement

Almost all optical communication depends in some way
upon muscular movement. Bioluminescent organisms may move
in space while emitting light or modulate emitted light by
means of movable covers (above). Other animals change
color by chromatophoric mechanisms involving muscular contraction. Animals that create extrinsic optical signals
such as the display bowers of bowerbirds do so by coordinated movements. Finally, animals use muscular movement
to change their body shapes, orient in space and make gestures.

Animal movement is due to subcellular microtubules
(ciliary and flagellar action) or myofibrils (muscle contraction). Anatomical details of muscular types, arrangements and innervation vary greatly among animals and likely

impose various constraints on signaling that have yet to
be studied. Vertebrate *smooth muscle*--responsible for
mammalian pupillary responses, vasomotor changes as in
blushing, and other optical signals--often has antagonistic sympathetic and parasympathetic innervation to the
same muscle cell. Some cells are polyterminally innervated (many junctions), but many are not innervated at
all, their activation being due to contraction of nearby
fibers, mechanical stretching or free neuro-endocrine
chemicals. Vertebrate smooth muscle is therefore likely
to be the basis of relatively slowly modulated signals.

Skeletal muscles tend to be arranged in *antagonistic pairs*, as when one muscle mass extends a limb and the
other flexes it. In crustaceans, which utilize limb movements as signals, muscle cells are innervated both polyterminally and polyneuronally (many different nerves),
allowing for precise and rapid control. Insect muscles
are similarly innervated, but lack peripheral inhibitory
innervation; instead, inhibition occurs centrally to prevent activation of excitatory motoneurons. This difference is probably important in the kinds of signal movements used by the two arthropod groups, but the subject
appears to be unstudied. Vertebrate *striated* skeletal
muscle has monoterminal and mononeuronal innervation, the
muscle cells themselves propagating an electrical spike
potential that activates the entire fiber more or less synchronously and at a constant rate under given mechanical
loading. It also appears as if there are slow- and fast-
contracting fibers in the same gross muscle. Contraction
of a single fiber is an all-or-none phenomenon, and about
10 to 1000 fibers innervated by the same nerve make a
motor unit. Contraction is graded because a muscle may
have hundreds of such units, only some of which are activated at a given time. Arthropod muscles are not organized into such units, their contraction velocity being
due to the polyneuronal innervation of individual fibers
that may contract slowly or rapidly. It seems possible
that the "jerkiness" of signal movements in arthropods
compared with vertebrate movements might be partially
traceable to these differences in muscular mechanisms.

Because of these control mechanism, the temporal modulation of signal movements is not limited by contraction
latency *per se* in vertebrate striated muscle. Hummingbird
wings may beat at 50 Hz, for example. Some insect wings
beat much faster--as high as 1000 Hz--a value that clearly

exceeds the capabilities of muscles to contract and relax with each wing-stroke. The elastic exoskeleton to which the wings are attached is bent by muscle contraction that loads the system with potential energy this is released when it pops back into shape--much like the way in which a metal soft-drink can pops back when the top is pushed down. The insect muscles then contract only occasionally to reinforce the oscillation--like pumping a child's swing on every third or fourth cycle. I have not found examples in which mechanical resonance systems are used in optical signaling, but clearly such rapid movement exceeds flicker-perception rates under most conditions *(ch 5)* and hence is unlikely to be useful in communication. It seems likely to conclude that muscle mechanisms impose no serious temporal problems for animal signaling.

biochromes

Coloration of animal parts is created by molecules that absorb light differentially within the spectrum (*bio-chromes*) and specializations of the integument that affect light through physical principles such as refraction, scattering and so on (*schemochromes*, below). The terms biochrome (D.L. Fox, 1953) or *zoochrome* (Needham, 1974) help prevent confusion due to the overworked word "pigment." In industry, a *pigment* is a coloring compound that is suspended in a base, such as pigments in paints, and a *dye* is a substance that complexes chemically with some substrate, such as dyes in cloth. Biologists, however, use the term pigment in a variety of ways, such as for the chromophore-protein complex in the eye that changes upon absorption of light (see visual pigments in *ch 5*) and for the chemical complexes of the integument that are photostable.

Biochromes in the integument are molecular complexes that absorb photons in particular spectral bands, turning the absorbed energy to heat *(ch 3)*, and reflect non-absorbed photons to give animals their observed coloration (see *eq 3.4*). Molecules that absorb uniformly in the visible spectrum, so as to reflect white, gray or black, are also classified as biochromes. The molecular factors that determine the absorption bands of biochromes include their large molecular size, bonding structure of component atoms, planar configuration, side-chains and rings, and conjugation with proteins and other materials.

Because of the many ways in which the molecules can differ, there are hundreds if not thousands of different animal biochromes. They tend to fall into about a dozen major classes based on chemical similarities *(table 4-I)*,

Table 4-I

Roughly Equivalent Classifications of Animal Biochromes by Different Authors

D.L. Fox (1953)	H.M. Fox and Vevers (1960)	Needham (1974)
carotenoids	carotenoids	carotenoids
quinones	quinones	ternary quinones
tetrapyrroles: porphins & bilins	haemochromogens, porphyrins & bilins	pyrroles
indoles: indigoids & melanins	melanin	indoles
"	sclerotein, ommochromes & Tyrian purple	ommochromes or phenoxazones, *etc.*
purines & pterins	guanine, pterins & flavins	other *N*-heterocyclics
flavins or lyochromes	"	"
anthocyans & flavones	(miscellany)	chromans
chromolipoids or lipofuscines	"	fuscins
(miscellaneous)	haemoglobin & chlorocruorin	metalloproteins
(miscellaneous)	(miscellany)	(other)

but most biochromes are incompletely characterized, so
that the classification is likely to change. Many biochromes were first named before anything was known of
their structure, then renamed to express general chemical
similarities with other known biochromes, then renamed
again when something of their chemical structure was understood. Even the major classes have various names, as
indicated by comparisons in *table 4-I*. The approximate,
if not complete, chemical structure is now known for at
least one biochromes in each major group.

Some biochromes are found principally in plants (*e.g.*,
flavins) and hence are not relevant to animal signals,
whereas others are rarely found in the integument (*e.g.*,
blood hemoglobin) and hence manifest coloration in other
ways, such as blushing when blood flows close to the skin.
Because of molecular variation, densities of pigmentation
and anatomical structures, the same group of biochromes
may give rise to various colors of the integument (*table
4-II*, p. 98). From the table it appears that carotenoids,
melanins and fuscins provide particularly favorable systems for creating different colors with minimum chemical
changes, and hence might be readily changed with minimum
genetic change in animal populations. Although I could
find no systematic reviews of biochromes used in animal
signal-colors, it appears from extensive literature that
melanins and carotenoids are the most common bases of
coloration in molluscs and bird feathers (also see below).

Melanins are polymers, often attached to a protein
as a granule, and *carotenoids* are long hydrocarbon molecules with conjugated double-bonds and a terminal cyclo-
hexene ring. In birds, melanins are readily synthesized
from tyrosine (an amino acid), but carotenoids must be
ingested as part of the diet. *Carotenes* are pure hydrocarbon carotenoids sometimes found in avian soft-parts
(legs, beaks, bare skin, eyes), and their oxidation yields
xanthophylls, the principal carotenoids of feathers and
soft-parts. Ingested carotenoids may be deposited directly or changed chemically before deposition as biochromes.

Inspection of *table 4-II* (next page) shows that whereas long-wavelength reflection may be created by various
biochromes, the violets, blues and greens are rarer. Ca-
rotenoid-protein complexes exist as blue biochromes, and
blues are often created by schemochromes as well (see next
section). The comparative rarity of extractable blue

Table 4-II

Biochrome basis of Animal Coloration
(compiled from various sources)

biochrome class or subclass (see *table 4-I*)	chromatic						brown & tan	neutral		
	violet & purple	blue	green	yellow	orange	red		black	gray	white
carotenoids	x	x	(x)	*	*	*	x	x	x	-
quinones	-	x	-	x	-	x	-	-	-	-
pyrroles	x	x	(x)	x	-	x	x	-	-	-
melanins	-	-	(x)	x	x	x	*	*	x	-
ommochromes	x	-	-	-	x	x	x	-	-	-
purines	-	-	-	-	-	-	-	-	-	*
pterins	-	-	-	x	x	x	-	-	-	x
flavins	-	-	-	x	-	-	-	-	-	-
fuscins	-	-	-	x	x	x	x	-	-	-
metalloproteins	x	-	(x)	-	-	-	x	-	-	-
other	x	-	*	x	x	x	x	-	-	-

* especially important, x occurs, (x) usually in conjunction with another biochrome or a schemochrome, - rare if occurs at all

dyes from animals made the Royal Tyrian Purple of the ancients particularly values (extracted from the shell of the gastropod *Murex*). Even today, blue and purple remain symbols of royalty.

Green-reflecting biochromes are so rare *(table 4-II)* that most animal green is created by schemochrome-biochrome combinations (see next section). Green is the sensation produced by light from the center of the visible spectrum so that a green biochrome must have two absorption bands: one at each end of the spectrum. This fact predisposes green coloration based on biochromes to show dichromatism *(ch 3)*. It may be that blue and green optical signals are relatively rare (especially among birds) because of the difficulty of evolving mechanisms to create these colors, although an obvious problem with green signals is that they would not contrast well with a background of green foliage (see *ch 7*). Green is a common color among some reptiles and insects, where it presumably functions in concealment *(ch 6)*.

Although the comparative research comprises a very scattered literature, it appears that virtually any class of biochrome may be expected in any major animal phylum. There are, however, certain emphases and restrictions in particular phyla, and especially in taxonomic classes. Of the animals that commonly communicate optically, cephalopods utilize ommochromes in the integument and melanin in their ink; arthropods have a diversity of pigments, including ommochromes and pterins, with porphyrins being rare; and vertebrates use most of the biochromes, especially melanin.

More specifically among arthropods, the crustaceans possess carotenoids, bilins, melanins, ommochromes, pterins, flavins and metallic-based pigments. Virtually every sort of pigment has been discovered in one insect or another. Vertebrate pigmentation is more restrictive, with bony fishes, amphibians and reptiles possessing primarily melanins, carotenoids, pterins and flavins; bird feathers relying primarily on melanins and carotenoids (also structual colors); and mammals being virtually restricted to melanins in hair and skin.

Despite the array of biochromes available to animals for signals, puzzling metabolic constraints on the evolution of such pigments exist. As A. Brush *(in lit.)* has pointed out, animals synthesize melanins but not carotenoids, and even within a class of animals such as birds, metabolic pathways of biochrome chemistry vary without apparent pattern. "Why aren't alternative pathways used? What determines the degree to which molecules can be modified by animals? Is it a matter of the presence of the

proper enzyme array, the energetics of synthesis or the
stability of the end product?" Future research in these
areas may yield valuable clues to understanding the coloration of animal signals.

A final point concerning biochromes seems particularly important. Because one class of biochrome may provide
various colors *(table 4-II)*, very small genetic shifts
can provide dramatic color differences among animals. Test
(1942) first pointed this out in flickers *(Colaptes)*, the
yellow-shafted and red-shafted forms being considered separate species until quite recently. The yellow under-wing
and under-tail coloration of the former becomes reddish
orange in the latter, and the male's mustache mark is
black in the former and red in the latter (there are other
subtle differences). It appears that the red-orange-yel-
low series in flickers *might* be due simply to concentration of biochrome. Brush (1970) found in a convincing
analysis of red and yellow coloration of tanagers *(Ramphocelus)* that different concentrations of the same carotenoids were responsible for the whole array of colors.
Such results suggest a two-part conclusion: coloration
may be evolutionarily labile within a single biochromic
system, yet in a more general sense evolutionarily constrained to that system once it is evolved.

schemochromes

Animal coloration could also be based on any of at
least five physical principles explained in *ch 3*: refraction, Mie scattering, Rayleigh scattering, diffractive
interference and thin-layer interference. The spectral
dependence of the index of refraction *(eq 3.8)* may be u-
tilized to select certain colors from complex bioluminescent organs containing transparent lenses, but in general
does not appear to be an important factor in animal coloration. Mie scattering by particles that are relatively
large compared with the wavelength of light is probably
an important factor in diffuse white coloring, but is not
known to have noticeable spectral effects on animal coloration. Most schemochromic coloration is therefore due to
Rayleigh scattering or interference.

Blue coloration of bird feathers usually appears to
be due to schemochromes rather than biochromes, but just
why the blue jay, for instance, is blue remains a matter
of dispute (see Brush, 1972). The classical explanation

is that minute melanin particles in the feathers are responsible for Rayleigh back-scattering of short wavelengths
being absorbed by a heavy melanin layer beneath. Dyke's
(*e.g.*, 1971) studies of parrot feathers, however, suggest
an entirely different mechanism in which hollow keratin
cylinders seem to be responsible for thin-layer interfer-
ence-reinforcement of short wavelengths. Some green
feathers, in any case, arise from a blue schemochrome
mechanism plus yellow biochrome.

Better understood than feather schemochromes are those
in scales of fishes due to layers composed of crystals of
guanine and hypoxanthine, two similar nitrogenous compounds.
The crystalline layers are aligned in stacks whose spacing
determines the spectral band reflected (see *fig 3-9*, p. 71)
according to thin-layer interference. The primarly manifestation of this schemochromic mechanism in fishes is the
nearly total reflection of visible light due to the overlapping of reflecting stacks with different spacing, providing most fishes with their mirror-like, silvery reflection.

Iridescence is the name given to metal-like reflectance of a whole range of colors, and is always due to interference. In molluscs, the iridescence in mother-of-
pearl is due to calcite layers that create primarily thin-
layer interference. It was once thought that non-iridescent blues could be attributed to Rayleigh scattering and
iridescent blues to interference, but as pointed out above
interference may give rise to non-iridescent colors as well.
In sum, Rayleigh scattering (often called *Tyndall scattering* in the literature on schemochromes) has been reported
for reptiles, birds and mammals, and in the latter is the
cause of blue eyes such as mine. Diffractive interference
appears to be rarer, but has been reported as the basis of
coloration in some aquatic invertebrates and some insects.
Thin-layer interference is the process most commonly attributed to schemochromes, being reported for a cephalopod,
various crustaceans, transparent insect wings, solid colors
of butterfly wings, beetles, fishes, reptiles and birds.

color-change

The rapidity with which animals change color varies,
and is governed by various mechanisms. Seasonal changes
in birds and mammals are due primarily to *molt* of the

plumage or pelage, respectively, although in some cases
it is due to wear of the feathers or hairs. A well-known
example of wear is the assumption in spring of the black
throat-patch of the male house sparrow as the gray tips
of the feathers wear away. Molt is often triggered by
changes in day-length and mediated through hormonal control.

More rapid, hormonally controlled color changes are
due to edema-like swellings, particularly in the genital
areas of many mammals that become pink due to surface
vascularization during the reproductive season (Hailman,
1977a: figure 30). In other cases, specific seasonal
biochromes are laid down under hormonal control, as in the
beak of the starling, which turns from black to yellow in
early spring.

The most rapid color changes seem always due to *chromatophores*: special structures that alter color through
dispersion or concentration of biochromes. Most animals
that communicate optically have evolved chromatophoric
capacities to some extent, including crustaceans and cephalopods (plus a few insects) among invertebrates, and fishes,
amphibians and reptiles among vertebrates. The covering
of inert feathers and fur make such mechanisms improbable
in birds and mammals.

There are three kinds of chromatophores. Cephalopods
possess *chromatophoric organs* with muscle fibers that can
act rapidly, the changing color of the Mediterranean octo-
pus having been known to Aristotle. Crustaceans have *chromatophoric syncytia* that appear to work by streaming action
of cytoplasm carrying the biochrome. Vertebrate chromatophores are *cellular*, and the color-changes of the famous
African chameleon were also known to Aristotle.

Chromatophores are frequently named for the biochromes
they contain or the coloration they create, so that melan-
ophores contain melanin and xanthophores produce yellow
coloration. Guanine-containing chromatophores are guano-
phores or leucophores, and those responsible for iridescent
sheen are iridophores or iridocytes. When the biochrome
is concentrated into a tight ball (*punctate* state) it lends
little to the animal's color, but when dispersed (*reticu-
late* state) covers any underlying biochrome and hence
changes the animal's color. Invertebrate chromatophores
may contain two or more pigments, but cellular vertebrate
chromatophores each contain a single pigment. Despite
their different cellular mechanisms, cephalopods and fishes
seem capable of altering their color faster and through a
greater range than any other animals.

Types and Origins of Signals

Animals use the mechanisms of light-modulation just
reviewed to create spatiotemporal arrays of photons that
constitute the signals of optical communication (ch 2).
The types of signals and their evolutionary origins are
reviewed with examples and illustrations in Hailman (1977a),
which is here summarized and extended in certain ways.
The detailed aspects of optical signals that actually encode information, which is to say the sign-vehicles (ch 2),
must be determined experimentally in each individual case
to discover which variables of the signal affect the receiver's behavior.

types of signals

Visual stimuli may be divided roughly into two broad
classes: (a) spatially unpatterned light that encodes information primarily by physical intensity and perhaps
spectral distribution, and (b) spatially patterned light
that gives rise, for example, to the visual images with
which we humans are introspectively familiar. Spatially
unpatterned ambient illumination is important in the control of such behavior as migratory restlessness in birds,
phototactic responses in many animals, circadian rhythms
and so on, and may be sensed with simple photoreceptor
organs such as ocelli. Spatially patterned light is more
important in animal communication, and requires more sophisticated photorecepotrs such as the compound eyes of arthropods or the image-forming, camera-like eyes of vertebrates.
    The first distinction in classifying optical signals
is whether the source of patterning is the sender itself
(intrinsic signals) or some object fashioned by the sender
(extrinsic signals). Extrinsic optical signals include
such objects as scratch-marks on a tree, the display bowers built by male bowerbirds to attract females (see figure 1 in Hailman, 1977a) and the type-symbols on this
page. Wilson (1975; 186) proposes that extrinsic signals
be called "sematectonic."
    Intrinsic optical signals are naturally divided into
those created by bioluminescence and those created by reflection of ambient light. The latter constitute the overwhelming majority of animal signals and have appropriately

received the most ethological attention. Reflected signals have both behavioral and morphological elements, and
the *behavioral elements* may be partitioned into three types
as shown in *fig 4-1*. The sender may assume a special *or-*

TYPES OF BEHAVIORAL SIGNALS

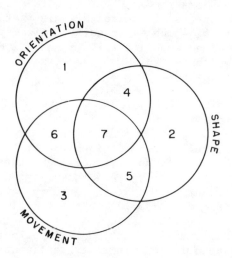

*Fig 4-1. Types of behavioral signals represented as a
Venn diagram of three elements of intrinsic, reflected-
light signals.*

ientation with respect to the receiver, may create a
special body *shape*, and may conduct specific *movement;*
all the combinations of these behavioral elements are
known from animals, as reviewed and illustrated in Hailman (1977a). Accompanying these behavioral elements are
morphological elements of signals that may be partitioned
into modifications of the reflecting surface itself and
larger structural aspects that alter the visible shape
of the reflecting surface, such as display plumes, antlers, *etc.* (see Hailman, 1977a for examples and illustrations).

The variety in signals due to specializations in the
reflecting surface are manifest in the *spectral reflectivity*, spatial *pattern* of reflectance and *type* of reflectance (diffuse *vs* specular; *fig 3-5*, p. 65). As dis-

cussed in the next chapter, receivers analyze spectral reflectivity according to sensations of hue, saturation and
brightness, together comprising the   perceived color of
the object viewed. Because the spectral reflectivity of
signals is rarely determined by physical measurement, e-
thologists tend to describe this variable according to
their own perception of color. In many cases it creates
little mischief to describe signals in terms of hue,
saturation and brightness *(ch 5)*, but it is important to
keep in mind that these are not physical descriptions.

The spatial variables in reflected-light, intrinsic
signals are summarized in *table 4-III*. For a discussion

Table 4-III

A Classification of Optical Signals
(based on Hailman, 1977a)

EXTRINSIC SIGNALS

INTRINSIC SIGNALS

  Bioluminescent Signals

  Reflected-light Signals

   BEHAVIORAL ELEMENTS

     Orientation

     Shape

     Movement

   MORPHOLOGICAL ELEMENTS

     Surfaces
       spectral reflectivity
       spatial pattern of reflectance
       diffuse vs specular reflectance

     Structures

---

of the specialized topic of bioluminescent signals the
reader should consult Lloyd (1977). I have not attempted
a classification of variables in extrinsic signals, which
are less common than intrinsic signals in animal communication, but many of the aspects of intrinsic signals given
in *table 4-III* apply to extrinsic signals as well.

phylogenetic origins

Animals may cue on any morphology or behavior of
other animals and hence such morphology and behavior constitute signals. However, this volume focuses on those
aspects of morphology and behavior that have been influenced during the course of evolution by selection pressures
to enhance their efficiency as social signals: ritualized
signals or displays (*fig 2-4*). All such ritualized signals are presumed to be evolved from non-signal behavior
(Darwin, 1872; Tinbergen, 1951) and therefore have *phylogenetic origins (ch 1)* that may prove crucial to identify
in order to understand signal characteristics. For example, the differences in threat signals of sparrows and
gulls can be most parsimoniously explained by differences
in the fighting methods from which the signals evolved
(Hailman, 1977a: figures 23 and 24).

The behavioral elements of optical signals have been
evolved from a huge range of non-signal behavioral patterns (Hailman, 1977a provides examples and illustrations).
Virtually any non-signal behavioral pattern may be changed
during phylogeny to become a signal, and this includes autonomic responses as well as movements mediated by skeletal musculature *(table 4-IV)*. In addition, certain optical signals have been evolved from behavior used to generate signals in other channels, a process I called *secondary ritualization* (Hailman, 1977a).

The phylogenetic origins of morphological elements
are not so well understood, and have been little studied.
In the final major section of this chapter, various selection pressures that affect morphological elements, such
as surface coloration, are discussed. It seems possible
that some coloration was originally evolved for purely
physical reasons, such as retardation of integument-wear
or protection from damaging ultraviolet radiation (next
major section), and other coloration was evolved for visual concealment from predators or prey *(ch 6)*. Then these
pre-exisiting capacities were ritualized into social signals.

Table 4-IV

Phylogenetic Origins of Behavioral Elements of Optical Displays, as Reported in the Literature (based on Hailman, 1977a)

INTENTION MOVEMENTS OF SKELETAL ACTION PATTERNS

Agonistic Behavior
    fighting and combat
    fleeing, flight and protective responses

Reproductive Behavior
    mounting and copulation
    penile-erection and lordosis
    nest-building
    parental care

Maintenance Activities
    rolling and wallowing
    scratching
    bill-wiping
    preening and grooming

Other Action Patterns
    locomotion
    foraging and feeding
    anti-predator behavior
    orientation of sense-organs

AUTONOMIC RESPONSES

Pilomotor Actions of Fur and Feathers
    sleeking
    fluffing and ruffling

Respiratory Responses
    yawning

Vasoresponses
    flushing
    blanching

Ocular Responses
    pupillary actions

"Pseudosignals"

The behavior and morphology of animals are shaped by all kinds of competing pressures of natural selection and hence it is misleading to search for a single adaptive function of each trait (see also Hailman, 1977b). In many cases, therefore, signal-like behavior or morphology may evolve because of selection having nothing to do with animal communication; I shall call such signal-like traits *pseudosignals*. It is important to identify pseudosignals as a way of re-emphasizing that the use of behavior and morphology for signaling must always be empirically established. However, a more compelling reason for this section is that characteristics of signals may represent compromises between selection for effective communication and selection for entirely different advantages to the sender; one must study the latter in order to understand fully the former.

behavioral stereotypy

Behavioral elements of intrinsic optical signals *(table 4-III)* tend to be highly similar from one performance to the next, except when variability itself is used to encode information. Behavioral stereotypy is not restricted to optical signals, however, so it is instructive to review other causes of stereotypy in orientation, shape and movement.

Specific orientation with respect to social companions may be used for receiving as well as sending signals. Animals with highly encephalized sensory receptors, such as terrestrial vertebrates, often turn their heads toward social companions to optimize sensory information about the appearance, sounds and smells of their conspecifics. In some cases, these simple orientational responses have themselves been the evolutionary origins of optical signales involving orientation, as noted previously.

Strikingly stereotyped postures or body shapes also look like optical signals, even when they are not. For example, male mammals of many species exhibit the characteristic flehmen posture after nuzzling the vaginal region of a female, which posture facilitates the chemical assessment of the female's fluids by a special organ in the roof of the male's mouth. Again, however, in certain species

the non-communicative posture has been the phylogenetic
origin of an optical signal (Hailman, 1977a). Other stereotyped postures have nothing to do with sensory perception
and are not known to be origins of optical signals. For
example, cormorants (Phalacrocoracidae) and anhingas (An-
hingidae) have feather structures that differ from those
of ducks and other waterbirds in that the feathers of the
former become waterlogged (Rijke, 1968). Therefore, these
birds often seek a perch in air after a prolonged period
of diving, spread their wings in a strikingly stereotyped
posture, and thereby dry out the feathers (*fig 4-2*).

*Fig 4-2. Wing-drying posture of the anhinga, a dramatic
and stereotyped pose that looks like an optical signal.
(From a photograph by the author.)*

Stereotyped movements are so common and widespread
among animals that these provide the greatest potential
source of confusion between signals and non-signal behavior. Any time an optimum manner of doing something
exists, natural selection may favor a stereotyped movement. Thus animals may move with a certain gait that
provides the most efficient locomotion, groom with fixed
movements that most efficiently clean the body surface
and so on. Stereotyped movements are therefore the least

diagnostic of behavioral elements in optical signals.

Finally, it is useful to emphasize again that behavior may appear to be generating optical signals when in fact the communication is in another channel. For example, tail-slapping of fish is a stereotyped, oriented movement that communicates by displacement waves in the water (Tinbergen, 1951), and release of communicative chemicals may involve stereotyped postures and movements (Hailman, 1977a: figure 13). Again, such behavior may secondarily give rise to optical signals *per se*.

biochromes and radiation-absorption

In the analysis of optical signals one is primarily concerned with the spectral reflectivity of animal surfaces *(table 4-III)*. However, biochromes act by absorbing light they do not reflect *(eq 3.4)* so it is possible that certain biochromes have evolved for their absorption rather than reflection properties. In such cases, the surface coloration of the animal may be a secondary outcome that is largely irrelevant to the primary function of the biochrome. There are two well-known phenomena of this type, and some suggested cases that have not been well studied.

Short-wavelength photons have relatively higher energetic content than those of longer wavelength *(eq 3.2)*, and ultraviolet radiation in particular is highly penetrating in biological tissues. Melanin absorbs radiation across a broad spectrum, and the dark peritoneal linings of lizards and possibly other desert animals are almost certainly evolved for UV-protection of internal organs (Porter, 1967). The extent to which animals have evolved melanic body covering for protection from ultraviolet radiation enjoys no consensus (Hamilton, 1973). It is possible that other biochromes confer similar advantages. Brush (1970) found cartenoid concentration in feathers of *Ramphocelus* tanagers to increase with altitude, where UV flux becomes greater, although he does not attribute the correlation to UV-shielding. Burtt (1977) found that UV penetrated white feathers most readily, carotenoid-bearing feathers significantly less, and melanin-containing feathers least of all.

There is little doubt that the heat arising from radiation absorbed by melanin at the surface of some animals is an important factor in their energy-balance (Porter *et*

al., 1973). However, there is doubt about how important
a factor this is in determining animal coloration (*e.g.,*
Hamilton, 1973). The role of melanin in avian feathers
and mammalian fur is particularly complicated, since reasonable but opposite arguments can be marshalled for expected effects. One may argue that black insulation would
transfer heat to the animal or that it would capture heat
at the periphery and reradiate the energy, thus keeping
the animal cooler. Porter *(pers. comm.)* has recent experimental evidence favoring the latter mechanism in mammalian hair, but there is also evidence for the former
action. For example, the Himalayan coat-color pattern
in domestic rabbits and Siamese cats consists of melanic
fur at the periphery of body projections such as pinnae,
nose, tail and paws. These distal tips are subject to
cooling to a greater extent than proximal areas of the
body. Reared in warm environments, the fur does not deposit melanin, but in cold rearing environments the coloration develops and then remains constant through successive
molts. Thus the Himalayan pattern seems adaptive for ac-
cumulating heat rather than dissipating it.

Burtt (1977) has recently provided convincing evidence that the nonfeathered legs of wood warblers (Parulidae) are colored in association with temperature extremes: dark-legged species stay north longer in the fall,
overwinter farther north and return north earlier in the
spring than do light-legged species. There seems little
question that melanin and possibly other biochromes are
involved in heat-balance of many animals, although just
how the mechanisms work may vary among species.

In addition to these two major absorptive functions
of biochromes may be added a few other suggestions. Loomis
(1967) points out that vitamin-D is synthesized by ultraviolet irradiation, and therefore melanic skin of man might
prevent oversynthesis in environments with high radiation
loads. Menaker *(pers. comm.)* suggested that the dark caps
of some birds might be effective in regulating the amount
of light that reaches photic receptors in the brain, which
are responsible for controlling daily and annual cycles of
rhythmicity (see Menaker, 1968). Finally, J. Baylis *(pers.
comm.)* points out that the black peritoneal lining of transparent bathypelagic fishes is thought to be an adaptation
to keep light *in*: the melanin prevents the bioluminescence
of prey in the fish's gut from being detected by other
animals.

biochromes and non-radiative protection

The deposition of biochromes in animals may provide forms of protection that have nothing to do with the ambient radiation on the animal. The primary example of this is the resistance to abrasion conferred by melanin deposition, but other examples have been reported as well.

Dwight (1900) appears to have first suggested that melanin in avian feathers retards wear, but this idea was supported only by arguments and anecdotes until Burtt (1977) recently conducted the first experimental demonstration of the phenomenon. Using a geological fossil-cutting gun under controlled conditions, he found that melanin-containing feathers showed much greater resistance to abrasion than did other feathers, whether or not they contained carotenoids. D.L. Fox (1962), however, suggests that carotenoids may also confer some advantage in abrasion-resistance. The mechanism of resistance to abrasion is not fully understood, but is probably not due to melanin *per se*. Needham (1974: 158) notes that melanin bonding to keratin strengthens the keratin, and melanin-containing feathers often show thickness surrounding the areas of deposition. It seems likely that at least some dark coloration in animals results specifically from selection for melanin-induced resistance to abrasion, as in the black wing-tips of otherwise white seabirds (Averill, 1923). Burtt (1977) has shown the highly abraded areas of wood warblers, such as the dorsum and the flight feathers, are all covered with melanin-bearing feathers, often colored by other biochromes as well.

Needham (1974) cites two other types of protection from biochromes in insects. Quinones tan insect exoskeletons, and it appears that dark *Drosophila* integument is less wettable than light-colored covering. Furthermore, the offensive odors produced by some arthropods appear to be due to p-benzoquinones from the exoskeletons. These odors, which offer protections from predators, are correlated with the chromatic properties of the molecules, so that coloration may be a by-product of the selection for chemical defenses. Lastly, some insects are known to deposit nitrogenous wastes in the exoskeleton as a method of excretion that prevents water-loss. Because the exoskeleton is molted once or twice a year, this form of protection against dehydration may give rise to particular external coloration as an irrelevant consequence.

In addition to these protective functions, biochromes
act to protect animals from predation through concealing
coloration and mimicry. For purposes of this volume, such
coloration is considered a form of optical communication
of "misinformation" and is accorded its own chapter *(ch 6)*.

biochromes, schemochromes and photoreception

Certain biochromes and schemochromes may be evolved
to aid visual processes, and these may be considered under
two general categories: specializations within the eye itself that have little effect on the surface color of the
animal, and specializations of the external covering that
are directly related to photoreception.

Three kinds of biochromes and at least one kind of
schemochrome are known from eyes of animals. In many invertebrate eyes there are *screening pigments* that affect
the spectrum of light allowed to penetrate to the photoreceptors. Somewhat similarly, oildroplets in the retina
and epithelium of vertebrate eyes may contain carotenoids
(Hailman, 1976c). The receptors themselves contain *visual
pigments* that change chemically upon absorption of light
as the primary process in photoreception *(ch 5)*. Finally,
all vertebrate eyes except those of albinos are lined with
a *pigment epithelium* containing melanin, the principal
function of which is to capture stray photons that penetrate the retina without being absorbed by visual pigments.
Functionally similar pigment is found in many invertebrate
eyes. In addition to these biochromes in the eye, the
pigment epithelium or retina itself in vertebrate eyes
may contain guanine plates similar to those in fish scales
that reflect light. Such *tapeta lucida* and analogous
structures function by thin-layer interference, and reflect photons to provide a second chance for their capture
by visual pigments during very low levels of illumination.
In high light levels the tapedial plates may be shielded
by extensions of the pigment epithelium. Tapedal reflection of light may be seen as *eyeshine* in nocturnal animals,
but eyeshine seems an artifact resulting from man's creation of highly directional light sources, and it is unlikely that eyeshine is a factor in animal communication
under natural conditions.

At least three photoreceptively related functions of
biochromes on the body surface have been proposed. Need-

ham (1974: 163-177) cites carotenoids and possibly porphyrins as bases of dermal light sensitivity, and states
(p. 177) that "probably in all animals there is a general
dermal light perception, whether or not discrete eyes also are present." Two possible uses of biochromes to aid
conventional photoreception concern coloration near the
eyes. Ficken (Ficken and Wilmot, 1968; Ficken *et al.*,
1971) has suggested that lines projecting anteriorly from
the eyes are used by birds and other animals to sight prey
for a feeding-strike. Burtt (1977) points out that such
lines are often dark and may instead be antiglare adaptations, although the two functions are not mutually exclusive. The antiglare strategy is also used to aid human vision, as when football players put lampblack on
their cheeks to diminish specular reflectance or when
airlines paint the metal below airplace cockpit windows
flat black. It seems likely that black around the eyes
could serve to diminish reflection in animals other than
birds (*e.g.*, the black mask of the raccoon), but a similar pattern of coloration would also be predicted as a
deceptive mechanism for hiding the eyes *(ch 6)*.

These various uses of biochromes and schemochromes
for other than signaling functions are summarized in *table 4-V*. It is unlikely that this list is complete, and
the table serves primarily as a reminder that signal coloration of senders may often be an evolutionary compromise
with other advantages conferred by biochromes.

Table 4-V

*Non-signal Uses of Biochromes and
Schemochromes Established or Suggested*

RADIATION-ABSORPTION	USES WITHIN THE EYE
UV-protection	screening pigments
thermal balance	visual pigments
vitamin-D synthesis	pigment epithelium
block bioluminescence	tapedal reflection
of prey in gut	
PHOTORECEPTIVE USES	NON-RADIATIVE PROTECTION
shield CNS receptors	abrasion-resistance
dermal sensitivity	water-proofing
sighting lines	chemical defenses
glare-reduction	anhydrous excretion

Overview

Animals encode information by modulating light
through bioluminescence, physical movement and selective
reflection. The mechanisms of modulation appear to impose no serious general constraints on the total amount
of information that may be sent, although in specific
cases there are certain limitations. For example, it may
be difficult to evolve a green-reflecting body surface
and when green surfaces do exist they may exhibit special
(not necessarily desirable) properties such as iridescence
or dichromatism. Limitations on the rapidity of muscle-
movement and color-changes probably set an upper limit on
the rate of informational transfer, but the limit appears
to be sufficiently high to be of minor consequence. An
ethological classification of signals follows naturally
from mechanisms of modulating light and emphasizes the
great variety of alternative signals potentially available to senders. The phylogenetic origins of signals may
dictate to some extent their characteristics, but the evolutionary sources for signals are so legion that they do
not limit the number of signals that may be evolved. Many
selection pressures, however, constrain the behavior and
morphology of animals, in some cases producing signal-like
characteristics ("pseudosignals") serving non-communicative
ends. Very likely, the characteristics of all signals are
compromises between efficiency of signaling and non-communicative functions.

Recommended Reading and Reference

Some of the relevant ethological material on optical
signals may be accessed through Hailman (1977a). Harvey's
(1952) standard book on bioluminescence was updated by a
chapter (Harvey, 1960) and a volume edited by Johnson and
Haneda (1966). See Lloyd (1977) for an overview of communication by bioluminescence. The mechanisms of movement
are covered in most comparative physiology texts, such as
Prosser and Brown (1961) or Hoar (1975). Coloration is
explained in D.L. Fox's (1953) standard work, slightly
revised in a new edition (1976), and more modernly in
Needham (1974). The book by H.M. Fox and Ververs (1960)
is written at a level for those without background in chemistry. Special volumes on carotenoids were provided by
Karrer and Jucker (1950) and Goodwin (1954). See also
Denton and Nicol (1965, 1966), Parker (1948), Fingerman (1963).

Chapter 5

THE RECEIVER

Seeing is believing. --Plautus (*ca.* 2000 BC)

After having traversed the channel, a signal arrives at a receiver *(fig 2-1)* where its information is extracted. The receiver's eye and nervous system act not only as filters of signals but also as processors of the information they encode. Therefore, the receiver can constrain how information is encoded in optical signals. This chapter concerns primary features of visual physiology and perception that prove useful to understanding the optical design of animal signals.

Achromatic Sensation

The primary carriers of information in visual signals are spatiotemporal arrays of photons that differ in total energy and frequency composition *(ch 2)*, which the eye perceives in terms of brightness, hue and saturation. Although all three variables are intimately related physically, physiologically and psychologically, it is useful to consider first achromatic sensation based solely on brightness and then to extend treatment to chromatic sensation involving hue and saturation as well.

transduction

The primary process in photoreception is the trans-

duction of electromagnetic energy to neural energy. This
complex process is achieved by large molecules of *visual
pigments* in the eye that absorb photons, changing their
electromagnetic energy to molecular energy that initiates
chemical changes in the molecule. The visual pigment molecule consists of a relatively small *chromophore* (light-
absorbing part, derived from vitamin A) attached to a
large protein moiety. The energy added by photons breaks
bonds holding the two parts together and the separation
(in vertebrate eyes) or partial separation (in invertebrate compound eyes) of the two parts initiates a chain
of complex chemical reactions that eventually give rise
to electrical signals in neurons of the visual system.

     Every known visual pigment molecule absorbs over a
broad spectral range, and its absorption spectrum resembles a probability-density distribution for reasons discussed in *ch 3*. Dartnall (1953) found that when absorption spectra were plotted by frequency rather than wavelength *(eq 3.1)*, all curves possessed nearly the same
shape, but were displaced on the frequency scale. Later,
Munz and Schwanzara (1967) showed that there were slight
but consistent differences in the shapes of absorption
spectra of pigments based on vitamin A_1 and A_2. In actuality, visual pigment absorption spectra measured in in-
tact photoreceptors by methods of microspectrophotometry
do not conform exactly to the shape of the two theoretical
spectra (*e.g.*, Liebman and Entine, 1968).

     If the receiver's eye contains photoreceptor cells
all having the same visual pigment, the eye's sensitivity
to light should be proportional to its probability of absorbing photons at a given spectral frequency. Therefore,
behavioral and physiological determinations of spectral
sensitivity should correlate with visual pigment absorption spectra, and where relevant comparisons have been
made the correlation is good (*e.g.*, *fig 5-1*). The absorption spectra of visual pigments, and hence the spectral
sensitivity of the eye containing them, are thought to be
evolutionarily adapted to the spectral distribution of
ambient light in the species' habitats (*e.g.*, McFarland
and Munz, 1975b).

     Pit-vipers and a few other animals transduce electro-
magnetic radiation in the infrared frequencies (*ch 3*) by
an entirely different mechanism. In pit-vipers, a mem-
brane covering a small pit absorbs IR quanta, thereby heat-
ing the pit and activating heat-sensitive receptor cells

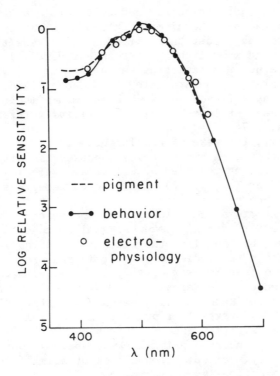

*Fig 5-1. Relative spectral sensitivity measured behavi-
orally and electrophysiologically with relative absorption of a visual pigment plotted for comparison (after
Blough, 1957). The data are expressed as logarithms of
the fractions of the maximum sensitivity or pigment extinction; behavioral data are from a tracking technique
using operant conditioning with the domestic pigeon; e-
lectrophysiological data from microelectrode recordings
in the pigeon's retina; and pigment data from the eye
of the domestic fowl.*

that line it. Such pit-organs cannot form images in the
manner of complex eyes, nor do animals use IR signals in
social communication. Reception in the near ultraviolet
frequencies by insects and some other animals is accomplished by visual pigments in complex eyes or simpler
ocelli, and hence such UV reception is true vision, albeit of frequencies invisible to us human animals.

As a footnote to the transduction process, it may be noted that the structural orientation of elements in photoreceptors may render them differentially sensitive to planes of polarized light (Waterman and Horch, 1966). More than a hundred invertebrate species are known to perceive the plane of polarization, and recent evidence suggests that some fishes, an amphibian, man himself and the pigeon possess at least marginal sensitivity (Kreithen and Keeton, 1974).

Finally, for sake of completeness it is useful to point out that even higher animals possess sensitivity to light that is not mediated by principal eyes or simple ocelli. Adler (*e.g.*, 1970) has studied receptors on the dorsal surface of the brain of amphibians that help control daily rhythms and orientational behavior. These receptors are structurally somewhat similar to eyes in the usual sense, but in birds the receptors controlling rhythms are known to be deep in the brain and presumably are not eye-like (see Menaker *et al.*, 1970). Furthermore, pigeon neonates have dermal sensitivity to light that does not involve the head at all (Heaton and Harth, 1974), and Needham (1974) states that probably all animals have dermal light sensitivity (see also Steven, 1963). In this volume, I am concerned primarily with stimuli perceived by the primary, paired, lateral eyes of animals that communicate optically.

brightness and contrast

When irradiance on the eye is increased, more photons are absorbed by the visual pigment molecules, and the visual sensation of light increases in magnitude. One way to measure the visual response is by placing one electrode on the surface of the eye and another elsewhere on the animal. When light is shined upon the eye, a complex electrical potential called the *electroretinogram* (ERG) is recorded between the electrodes, and the magnitude of this ERG measures (approximately) the visual sensation of *brightness*. The experiment yields a sigmoid function *(fig 5-2)* when only one kind of visual pigment molecule is activated by light.

The oldest problem in psychophysics is the exact nature of the function that relates visual response or sensation (ψ) to the physical intensity (irradiance, I).

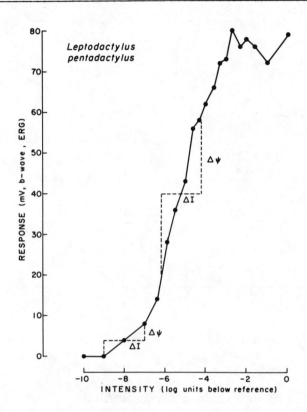

Fig 5-2. Intensity-response function of an ERG from the South American bullfrog (after Sustare, 1976). The response measure increases according to a sigmoid function of log intensity, so that near the center of the curve a given difference in log intensity between two stimuli (ΔI) causes a large difference in response (Δψ), whereas near the extreme parts of the curve the same ΔI causes only a small change in response.

There is agreement that

$$\psi \underline{\uparrow} I ,\qquad\qquad (5.1)$$

but the function is never linear ($\psi \neq t + kI$, where t is a threshold and k a proportionality constant). The Weber-Fechner relation postulates that $\psi \propto \log I$ and the Stevens power relation asserts that $\log \psi \propto \log I$; both relations approximate only portions of the sigmoid curve *(fig 5-3)*.

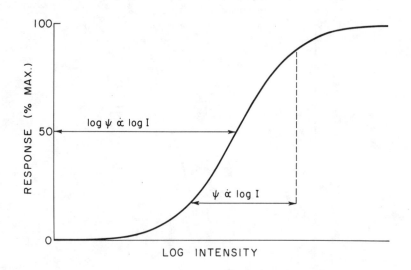

Fig 5-3. Idealized intensity-response function like that of fig 5-2. The middle part of the curve shows a nearly linear increase in response with log intensity (Weber-Fechner "law"), whereas the first part of the curve resembles a power function (Stevens "law").

Hailman and Jaeger (1976) point out that the empirical curves such as that of *fig 5-2* fit Crozier's *(e.g.,* 1940) model of the integral of a log-normal probability-density function; the extact nature of the function, however, is not of critical concern here.

It appears that the *brightness contrast* between two stimuli should be proportional to the difference in their brightnesses. Therefore, the perceived difference between two stimuli of different radiance depends not only upon the radiance-difference itself, but also on where this difference occurs on the intensity scale. Specifically, a given difference (ΔI) will evoke the greatest difference in brightness sensation ($\Delta \psi$) near the inflection point of the sigmoid curve *(fig 5-2)*. This implication is important for understanding visual adaptation considered below.

The perception of brightness is actually very complex,

and entire volumes are devoted to the subject (*e.g.*, Hurvich and Jameson, 1966). One principal phenomenon is of particular interest in optical signals of animals. When the eye perceives two adjacent stimuli of contrasting brightness, the difference in brightness is accentuated at the mutual border. Such *lateral inhibition* tends to emphasize the outline of a dark object against a lighter background (or *vice versa*), so that visual shape is enhanced by sharp borders of contrast *(ch 7)*. More is said about lateral inhibition near the end of this chapter.

An important perceptual phenomenon occurs when very bright light is in the visual field. Such light enters the vertebrate, camera-like eye not only through the pupil, but apparently partly straight through surrounding parts, and then is refracted and scattered by the ocular media to cause a general haze of light within the eye (Minnaert, 1954). This phenomenon, called *dazzling*, obscures vision in the field surrounding the source, decreases the sensitivity of the eye, and may cause momentary giddiness or pain. An apparent use of a dazzling signal is discussed in *ch 6*.

adaptation and duplexity

When the eye has been in the dark it becomes more sensitive to light (*dark-adaptation*) and when it bright conditions becomes less sensitive (*light-adaptation*). We are familiar with this phenomenon of adaptational shifts over large ranges of ambient irradiance, but it also occurs over small ranges as well. The effect of adaptation is to shift the sigmoid intensity-response curve *(figs 5-2 and 5-3)* toward the left (dark-adaptation) or right (light-adaptation) on its intensity axis. The usefulness of this shift is evident: by adapting such that the inflection-point of the curve is near the average ambient intensity, the eye maximizes the contrast between stimuli that deviate from the average (Hailman and Jaeger, 1976).

One can measure the time-course of dark-adaptation by determining the minimum intensity of light perceived by an animal (threshold) as a function of its time in the dark. For example, one may simply project lights of various intensities and ask a human observer whether he or she can see them. Animals can be trained by operant conditioning techniques to answer the same question by responding in

one way when they see light and in another way when they
do not. Therefore, an animal tracks its visual threshold
in such behavioral experiments, and one expects the threshold to decrease smoothly as a function of time in the
dark. For some animals, smooth dark-adaptation curves
result, but for many species there is a sharp break somewhere in the curve *(fig 5-4)*. This break suggests that

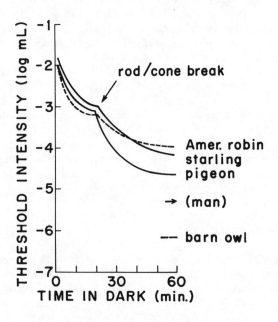

*Fig 5-4. Dark-adaptation curves for several species. As
a function of time in the dark, the eye becomes more sensitive (threshold falls). The break in the curve is attributed to differences between rod and cone vision in du-
plex retinae: the cone sensitivity falls to an asymptotic
threshold, but as dark-adaptation in rods continues the
visual threshold falls to still lower values until rod
sensitivity also reaches an asymptotic value. Fully dark-
adapted values for man and an owl are included for comparison.*

there are two populations of receptors having different
characteristics, and in vertebrate retinas these populations are known as *rods* and *cones*, named by the shape of

of their outer segments (portions that contain the visual pigments).

Experiments establish that rods are solely responsible for perception in the dimmest part of the intensity range (lower portions of curves in *fig 5-4)* and primarily cones for the brightest part. Cones also mediate color perception, but for the present that complicated subject may be deferred; color vision is not known to occur at very dim intensities mediated by rod vision alone. Eyes with two populations of photoreceptors such as rods and cones show visual *duplexity*, and such eyes are found among many invertebrates as well as vertebrates.

During the evening, when the ambient light falls from levels of cone-dominated to rod-mediated vision, the brightnesses of objects relative to one another change. Purkinje noticed this shift in the apparent brightnesses of flowers in his garden at sunset, and this *Purkinje shift* implies that the spectral sensitivity of the eye changes. Using an operant conditioning method similar to that used for determining the time-course of dark-adaptation, it is possible to measure the spectral thresholds of animals under high and low light levels. As expected, the former curve of *photopic thresholds* occurs at much higher thresholds than the latter curve for *scotopic thresholds (fig 5-5)*. The Purkinje shift is shown more clearly by inverting the threshold scale to create curves of *spectral sensitivity*, and expressing both curves in terms of their maxima *(fig 5-6*, p. 127): curves of *relative* spectral sensitivity. From such curves it is apparent that under scotopic conditions a light of 600 THz will appear brighter than a light of the same physical intensity at 500 THz, but under photopic conditions, equally intense lights of these two frequencies will have the reverse brightness relations.

photometry

Because every known eye is differentially sensitive across its visible spectrum according to one or more curves of spectral sensitivity *(e.g., fig 5-6)*, stimuli of the same total physical irradiance may evoke different visual brightnesses when they differ in spectral composition. It therefore proves useful to devise a measuring scale that incorporates spectral sensitivity such that two stimuli

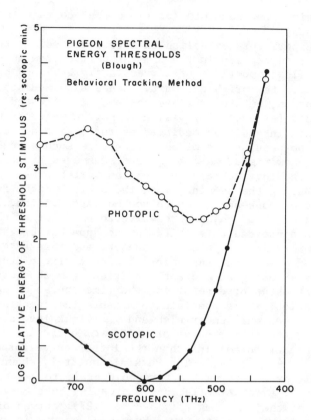

Fig 5-5. Scotopic and photopic spectral thresholds of the pigeon (after Blough, 1957). Light of higher energy is needed to stimulate the cones of photopic vision than the rods of scotopic vision.

that appear equally bright have the same value on the scale. Such scales have been devised for human vision, and they are the primary basis for the science of *photometry*.

The Commission Internationale de l'Eclairage (C.I.E.) has established two curves of standardized visual efficiency that are very close to the empirically measured scotopic and photopic spectral sensitivity curves of the human eye. Because photometry is concerned primarily with vision at high light levels, the C.I.E. photopic curve is

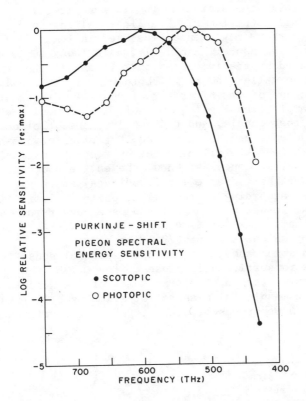

Fig 5-6. Purkinje-shift in spectral sensitivities, based on the data of fig 5-5 (opposite), plotted as log relative sensitivity curves. Each curve is normalized to its own maximum sensitivity to show more clearly the shift in spectral placement of the curves.

used as the basis of defining *photometric units* of measurement. Just as the distinction is made between the radiometric concepts of radiance and irradiance *(ch 3)*, photometry distinguishes the photometric concepts of *luminance* and *illuminance*, respectively, as the quantities of standardized brightness sensation.

Photometry has is own standard of *luminous intensity* called the *candela* (cd), or new candle, based originally on the apparent brightness of an actual, burning candle. A standard point-source candela emits a given luminous

flux through one steradian (solid angular unit), the flux
being called the *lumen*. The light falling upon an illuminated surface may then be measured in lumens/m^2, the unit
lux of illuminance. Similarly, a luminous surface may be
designated by the lumens/m^2 that it emits, the unit of
luminance called the *nit*. Photometry is initially confusing because of its diversity of units in different systems.
For example, the English unit of illuminance often used
is the foot-candle, and that of luminance the foot-lambert.

Radiance can be converted to equivalent luminance,
and irradiance to illuminance, by two mathematical conver-
sions. First, one must know the entire spectral distribution of the radiance (or irradiance) so that it can be
adjusted according to the C.I.E. photopic luminosity curve.
Then the result must be related by a conversion factor curiously named the *least mechanical equivalent of light*, the
value of which is about 679 lumens/W.

In some cases, the photopic spectral sensitivity of
animals is sufficiently close to the C.I.E. photopic luminosity curve that photometric units of luminance and illuminance are meaningful measures of stimulus brightnesses *(fig
5-7)*. In other cases, such use of photometric quantities

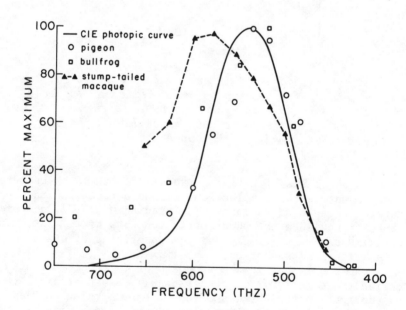

would be misleading, and photometric systems for the animal concerned should be devised.

Finally, confusion can arise over the use of names
applied to light. Heretofore, "intensity" has been used
to mean either radiance or irradiance, and it is also
used this way in the literature for either luminance or
illuminance. The photometric standard (candela) is a unit
of the quantity "luminous intensity" often abbreviated
simply to "intensity." Therefore, context should specify
which meaning of intensity is intended, and for most uses
one must further distinguish radiometric from photometric
quantities, as well as light emitted from a surface from
that falling upon a surface. Lastly, "brightness" has
been used to mean the subjective (photometric) intensity
of light, but "surface brightness" refers specifically to
luminance.

Chromatic Sensation

Color-blind animals and those communicating in very
low light levels receive optical information solely in
spatiotemporal arrays of brightnesses, but for most species
with well-developed optical communication, color plays an
important informational role. This section reviews some
aspects of color vision relevant to reception of optical
signals.

receptor-basis

If an eye has only one kind of receptor activated by
light under given conditions (*e.g.*, *fig 5-1*, p. 119), it
cannot distinguish two frequencies of light. For example,
a dim light at 550 THz may have the same brightness as an

*Fig 5-7 (opposite). Photopic spectral sensitivity curves
of some terrestrial vertebrates compared with the C.I.E.
photopic luminosity curve, upon which photometric units of
luminance and illuminance are based. The curves of the
pigeon (after Blough, 1957) and bullfrog (after Sustare,
1976) are reasonably close to the C.I.E. standard, but the
curve of the stump-tailed macaque (after Schierer and
Blough, 1966) differs markedly.*

intense light at 450 THz. If two receptor-types having different spectral sensitivity curves (e.g., fig 5-6) are simultaneously stimulated, however, certain judgments of frequencies independent of brightness might be possible. Scotopic and photopic systems are not ordinarily used together for such purposes; they are systems designed instead to operate at different ranges of light-level, and hence extend the sensitivity range of the eye.

Suppose for a moment that the two curves of fig 5-6 could be used simultaneously, and that their absolute as well as relative sensitivities were the same as their peak responses. They might be two types of cones having different visual pigments. In this hypothetical case, a stimulus at 450 THz would stimulate the two populations of receptors quite differently, but one at 550 THz would cause about equal activation of the two systems (p. 127). Therefore, by comparing activation of the two populations of receptors, the eye could make some primitive spectral discriminations. Such a two-receptor system is called dichromatic color-vision (not to be confused with dichromatism of colored substances mentioned in chs 3 and 4).

The difficulty with a dichromatic system is that certain frequencies cannot be distinguished. For example, in fig 5-8a a light of reference frequency r that stimulates both receptors in some ratio could be mimicked by a combination of two other lights, each of which stimulated only one (or primarily one) receptor by virtue of being outside the spectral range of sensitivity of the other (fig 5-8b). By independetly adjusting the intensities of these two superimposed frequencies a perfect metameric match to frequency r would theoretically be possible.

If three or more populations of receptors having different spectral sensitivities were used, confusions become more difficult to create. However, the neural coding required to compare the responses of the different populations and extract from the comparisons the frequency-composition of the stimulus become correspondingly complex. It is therefore of little surprise that most animal color-vision systems known utilize three or four different receptor populations as the best compromise. In the eye of the goldfish, for example, there are three different populations of cone photoreceptor cells (fig 5-9) that make up the trichromatic color-vision system, and similar systems are known in such species as the honeybee, rhesus macaque and human animal.

Fig 5-8. Equivoca-
tion in a dichromat-
ic system. An ani-
mal having two re-
ceptor-types in its
eye would confuse
a reference stimu-
lus of frequency r
with a combination
of two other stim-
uli (i and ii)
whose intensities
were properly ad-
justed to create a
metameric match.
Such equivocation
is more difficult
to achieve with
three or more over-
lapping sensitivity
curves.

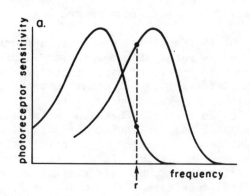

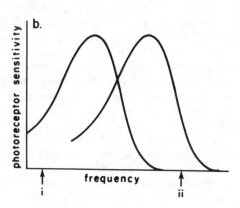

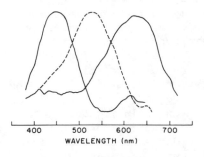

Fig 5-9. Trichromatic system of three cone pigments
determined by microspectrophotometry of the goldfish
retina (after Marks, 1965).

spectral discrimination and hues

If the visual system is extracting spectral information by means of comparing the activity of different receptor-types, one expects the precision of information to be different in different parts of the spectrum. The precision may be measured by the minimum detectable difference in frequency between two stimuli, called in psychophysics the *just-noticeable difference* (jnd) or *difference threshold*. The jnd of discrimination problems is never an absolute value: it is rather always based on some criterion of reliable discrimination, such as detecting the difference nine times in 10, or 99 times in 100 trials. When the same criterion of discrimination is used for experiments throughout the spectrum, then the relative jnd values may be plotted as a *spectral discrimination curve*.

From the trichromatic nature of many color vision systems *(e.g., fig 5-9)* one may guess at the general shape of the spectral discrimination curve to be expected. In the spectral region of high sensitivity of one receptor *(i.e.,* its peak) and low sensitivity of the other two, one expects poor precision because small changes in frequency cause only small changes in ratios of activity among receptor-types. In regions where receptors are more similar in sensitivity (on the slopes of their sensitivity curves) and sensitivity changes greatly with small changes in frequency *(i.e.,* the slopes of the sensitivity curves), one expects good precision in resolving frequencies. Therefore, *fig 5-9* (p. 131) suggests that spectral discrimination curves will be roughly shaped like a "W," rising rapidly toward large jnd values at the ends of the spectrum where one receptor is primarily active, falling toward small jnd values more centrally in the spectrum, but having a small hump in the middle where one receptor-type has its peak. Empirically determined spectral discrimination curves have precisely this expected shape *(fig 5-10)*.

From the shape of the spectral discrimination curve one may derive other expectations concerning the subjective impression of different frequencies. Where discrimination is poor (peaks of the curve in *fig 5-10)*, the animal confuses adjacent stimuli readily, so that these must appear subjectively similar. One expects, therefore, to experience three quite different sensations corresponding to the three spectral regions of large jnd values: the *primary* color sensations. We humans cannot ask animals

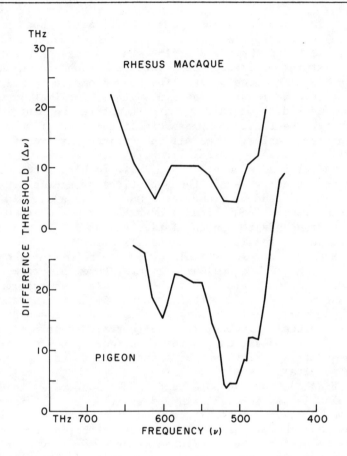

Fig 5-10. Spectral discrimination curves for two species (after Hailman, 1967b). Both curves show the general "W" shape expected from a trichromatic system of color vision.

what they experience, but we each know that for us high frequencies (short wavelengths) evoke violet and blue sensations, middle frequencies green and yellow sensations, and low frequencies (long wavelengths) orange and red sensations. There is as yet no precise mathematical model that relates such *hue names* of human sensation to the human spectral discrimination curve, but the generally expected relationship exists. Because we can make much finer spectral discriminations than indicated by the spectral

blocks represented by hue names, the number of hues a human culture recognizes linguistically is somewhat arbitrary, although their underlying sensory bases are not.

Mixtures of frequencies evoke sensations to which we
apply the same names as used for monochromatic stimuli,
even though we readily distinguish them. In fact, monochromatic radiation is very rare in nature; we experience
primarily spectrally complex stimuli composed of many different frequencies. When all the frequencies are of approximately equal physical intensity, the sensation is
achromatic: shades of gray from white to black that differ
only in intensity, as in the sensations of scotopic vision.
When a broad band of frequencies makes up the stimulus,
we identify it linguistically by a hue name such as blue
that corresponds to monochromatic hues at the frequency
of the peak intensity of the complex stimulus.

However, complex stimuli appear to be mixtures of
a monochromatic hue and white light. Pure, almost monochromatic stimuli are said to be *saturated*, whereas those
that seem to be mixed with achromatic light are *desaturated*. In sum, we distinguish three elements of sensation
about a visual stimulus: its hue (correlated primarily
with spectral position), saturation (correlated primarily
with spectral purity) and brightness (correlated primarily
with physical intensity).

Finally, one can note that when a spectrally complex
stimulus is double-peaked in its spectral energy distribution it may be difficult to predict what hue sensation
it evokes: the primary sensation may be that associated
with one or the other of its peaks or some mixture of sensation, or even some achromatic sensation (particularly
if the brightness is low). One special kind of double-
peaked stimulus evokes an entirely new hue sensation not
matched by any monochromatic stimulus: mixtures of the
spectral extremes (red and violet or blue) evoke the sensation *purple*, which is a uniquely non-spectral chromatic
hue. It is easy to see why purple is a unique sensation:
purple lights stimulate primarily the two extreme receptors *(fig 5-9)* and hence have a physiological effect different from that of any single monochromatic light.

metameric matching

Figure 5-8 (p. 131) presented a case of chromatic

equivocation in which two stimuli that have different physical bases (frequency compositions) are confused by the eye. Indeed, the whole point of utilizing three receptor-populations with different spectral sensitivities is to reduce such chromatic equivocation. Yet, in well-defined situations equivocation still occurs as *metameric matches* of stimuli.

Suppose one stimulus (monochromatic or spectrally complex) is shown on a screen adjacent to another screen illuminated by three superimposed lights that form the second stimulus. With only a few well-defined exceptions that are not of present concern, the three superimposed lights may be monochromatic or complex lights and may be chosen at will from all possible lights. It turns out that if the intensities of the three lights are independently adjusted, one can always create a visual match between the two screens.

This counter-intuitive result is subject to only minor specification, in that in order to create some matches one of the three lights must be superimposed with the reference stimulus of the first screen rather than with the other two experimental stimuli. The reason we do not experience such chromatic equivocation all the time in our daily lives is that the three matching lights must have just the right combination of intensities, which is sufficiently improbable to make metameric matching more important as an experimental tool for studying vision than it is a source of serious visual confusion. We do often see two surfaces, such as dyed clothing and painted wood, as having the same color when their spectral compositions may be very different.

Goldsmith (1961) devised a tentative colorimetric system for the honey bee based on matching experiments performed by Daumer (1956). Because the honey bee is sensitive to ultraviolet radiation, Goldsmith chose as theoretical matching stimuli one monochromatic stimulus in the UV (at 360 nm); the other two chosen lie within our visible spectrum as well as that of the bee (440 and 588 nm). Any stimulus may be plotted according to the percentage of its component energies at these three wavelengths, and since two percentages determine the third by subtraction the plot may be made on ordinary Cartesian coordinates *(fig 5-11)*.

This *chromaticity diagram* of the bee is read according to the following example: a monochromatic stimulus of

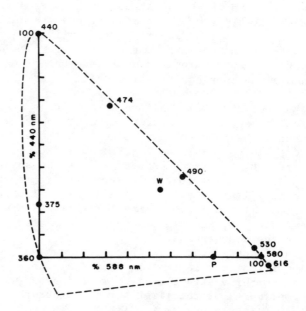

Fig 5-11. A chromaticity diagram of the honey bee (after Goldsmith, 1961). A monochromatic light of wavelength 474 nm is confused by the bee with a light having about 30% 588-nm and 70% 440-nm components. Some stimuli cannot be matched with combinations of 588- and 440-nm light, and require in addition some UV component of 360 nm, chosen as the third reference wavelengths for this particular diagram. There are an indefinitely large number of such possible diagrams, using various combinations of three reference stimuli.

490 nm may be matched by a combination of about 65% 588-nm light (x-axis) and 35% 440-nm light (y-axis), with little or no 360-nm light required. The broken curve in *fig 5-11* is the *spectral locus* of monochromatic lights, within which all real stimuli plot as points. Notice that the white-point (labeled *W* in *fig 5-11*) is matched by about 55% 588-nm light, about 30% 440-nm light and therefore requires about 15% 360-nm light for a match. In other words, a-chromatic stimuli of the bee require some ultraviolet components of energy. Flowers that appear white to us because they reflect about equal amounts of energy through

our visible spectrum may or may not appear achromatic to
bees, depending upon whether the stimuli do or do not reflect a sufficient UV component.

colorimetry and stimulus-space

The *C.I.E. chromaticity diagram* for standardized human color-vision is more complexly derived than Goldsmith's
diagram for the honey bee, but has the same major features.
The standard matching stimuli are mathematically defined
complex lights that are purely imaginary. Data from any
matching experiment, however, may be transformed mathematically for plotting on the C.I.E. diagram. Suppose a
complex light plots at point *P* in *fig 5-12*. A line from

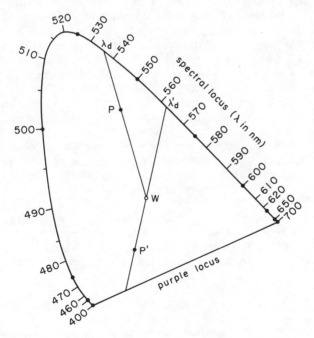

*Fig 5-12. The standard C.I.E. chromaticity diagram, with
the axes omitted. The white-point (W) is equal energy.
Every real stimulus plots as a point (e.g., P) and a line
from W through P intersects the spectral locus at the
dominant wavelength (λ_d). The ends of the spectral locus
are connected by a line called the purple locus. Purple
stimuli (P') are designated as explained in the text.*

the white-point W through P intersects the spectral locus
at point λ_d, which is the *dominant wavelength* of the stimulus. Because the human eye sees various similar stimuli
as achromatic, the dominant wavelength has meaning only in
relation to a given white, the white in *fig 5-12* being that
of equal energy throughout the visible spectrum. The ratio of the lengths WP to $W\lambda_d$ is the *excitation purity*, a
measure of the saturation of the complex light. Excitation purity varies from zero (white light) to one (monochromatic light).

When a stimulus plots in the lower portion of the
C.I.E. chromaticity diagram, the line from white (W)
through its point (P') may intersect the straight line at
the bottom of the diagram connecting the extremes of the
spectral locus (*fig 5-12*, previous page). This line is
called the *purple locus*, and since purples have no monochromatic referents they are designated differently from
other stimuli. A line is drawn from P' through W to intersect the spectral locus, and the point of intersection
is the *complementary dominant wavelength*, λ'_d. Whereas the
dominant wavelength is that monochromatic stimulus added
to white to produce a visual match with a given light, the
complementary dominant wavelength is that monochromatic
light to be substracted from white to produce a visual
match with a purple stimulus.

By means of tolerably straightforward calculations,
the chromaticity coordinates of any stimulus may be calculated for plotting on the C.I.E. diagram. One may then
graphically determine the dominant wavelength (or complementary wavelength) and excitation purity of the stimulus.
If the luminance of the stimulus is also calculated, any
stimulus may be expressed by three numbers. *Table 5-I*
shows the relations among physical variables, psychophysi-
cal specification and subjective sensations of stimuli.

Table 5-I

Variables of Surface Coloration of a Stimulus

psychophysical quantity	subjective sensation	physical correlate
dominant wavelength	hue	spectral peak
excitation purity	saturation	spectral variance
luminance	brightness	radiance

A number of different schemata have been devised for plotting any stimulus in a three-dimensional space. A painter named *Munsell* devised a subjective system based on variables he called hue, chroma and value, which are roughly related to dominant wavelength, excitation purity and luminance, respectively. The *Ostwald* system is similar, except based explicitly on the C.I.E. variables of colorimetry. It may be pictured spatially as two cones base-to-base, with the upper apex being pure white and the lower one being black, the axis between them representing the achromatic locus. The distance laterally from the central axis is saturation and the surface of the figure is the locus of monochromatic radiation. A section normal to the axis roughly resembles the C.I.E. chromaticity diagram. The ornithologist *Ridgway* devised a similar system (with variables he called color, tint and shade) that was used to describe the coloration of bird species in hundreds if not thousands of technical papers. E.H. Burtt, Jr. and I devised an Ostwald-like system derived from C.I.E. variables in which dominant wavelength, excitation purity and relative luminance were plotted as rectilinear variables in three dimensions, thus providing a cube in which all stimuli plot as single points (Burtt, 1977). Many similar systems have been utilized.

Perception

Optical signals encode information in spatiotemporal arrays whose component parts may be describable in terms of surface hue, saturation and brightness. However, the way in which the components combine to make larger assemblages perceived as whole must also be scrutinized: the spatiotemporal arrays are perceived as patterns, flicker, movement and so on. These higher levels of sensory processing are, in general, not precisely understood physiologically, but many have been characterized psychophysically to an extent that they prove useful in understanding the optical design of social signals.

temporal phenomena

Various simple temporal phenomena in vision probably imply limitations of the visual system in receiving and

processing stimuli. If one stares for a moment at a dark
shape on a light background and then looks at a homogeneous
dark background she sees the same shape as a light figure
on the dark background: the *negative after-image*. Reversed
polarity of the contrast yields reverse polarity of the
negative after-image, which is due to fatigue processes in
the visual system. Sometimes, under certain conditions,
the same polarity of contrast occurs in the after-image,
in which case it is a *positive after-image*, the causes of
which are not well understood. Furthermore, if the shape
or its background are chromatic, the negative after-image
may also be chromatic, but of the complementary color.

When some stimulus is rapidly alternated between dim
and bright states, one perceives the alternation as *flicker*
when the light is bright, but as a steady light (*fusion*)
when it is dim. The flash-rate at which the transition
occurs from flicker to fusion is the *critical fusion frequency* (cff), and in a variety of animals the cff as a
function of log intensity describes a sigmoid curve (*fig
5-13*). Presumably, the fusion results from slow decay
processes of the visual system causing a persistence of
sensation in the absence of stimulation, not unlike positive after-images in general effect.

Suppose a light is turned on in the left part of the
visual field, and then turned off just before another light
is turned on in the right part of the field. When the angular distance between lights is not too great, their brightnesses are similar and the extinguishing of the first light
is closely followed by lighting of the second, one perceives
a single light stimulus that seems to move from left to
right. This *phi phenomenon* shows that the perception of
movement may be related intimately to the perception of
flicker with a spatial component. Indeed, the experiments
upon which *fig 5-13* are based did not use flickering light
at all. Rather, the animals were placed inside rotating
drums with alternating black and white bars. When the
speed of the drum's rotation is low or the light is bright,
the animal moves either its eyes (*visual nystagmus*) or entire head (*optokinetic* or *optomotor response*) in an attempt
to track the moving stripes. However, the animal can turn
only so far without moving bodily, so when reaching the end
of its arc, snaps back to the forward orientation and immediately tracks a new moving stripe. When the speed of
rotation is great or the light is dim, the animal shows no
such responses because the stripes are visually fused and

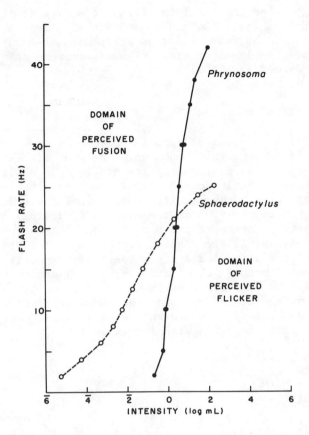

Fig 5-13. Flicker fusion curves of two reptiles (after Bartley, 1951). At flash rates above the plotted critical fusion frequency (cff) for a given intensity the animal perceives a steady light (fusion) instead of a flickering one.

the animal cannot perceive any motion of the drum.

Black and white stripes on a rotating drum (or sectors on a spinning wheel) fuse to an achromatic color of some intermediate intensity. If the stripes or sectors are of different hues, these colors fuse to either achromatic or chromatic colors, depending on conditions. For example, the apparent color of a wheel composed of a red and a yellow semicircle fuses to orange when rapidly

spinning. When three colors widely separated in the spectrum (such as red, blue and green) are used in the right
proportions, the fused color is achromatic, usually perceived as some level of gray. These phenomena may be
termed *movement-fusions*.

simple spatial phenomena

In some ways, simple spatial phenomena of vision resemble simple temporal phenomena. The spatial resolution
of the eye may be measured by its *acuity*, the ability to
discriminate closely spaced stimuli such as lines ruled
on a glass plate. The acuity of an animal's eye depends
upon many factors, such as the dioptic (focusing) system,
the density of photoreceptors and so on. Like temporal
resolution, spatial resolution depends upon the brightnesses of the stimuli and their spectral compositions, but
experimental methods of determining visual acuity vary
widely so that direct comparisons are difficult. *Figure
5-14* shows an acuity curve for the honey bee and two curves
for man determined by different methods. All three curves
show a sigmoid relation with log intensity.

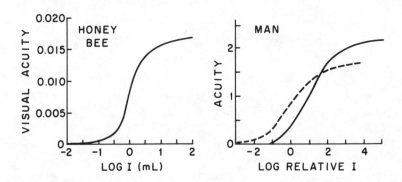

*Fig 5-14. Visual acuity curves of the honey bee and man
as a function of log intensity. Acuity is expressed as
1/(resolvable angle in minutes of arc) so that the maximum resolution of the bee is about 60' or 1° (after Pirenne, 1967) and that of man is about 0.5' (after Bartley,
1951). The two curves for man show results of two testing
methods.*

Visual acuity is not a single, invariant function of intensity, however. For example, acuity of amphibious animals in air and water may be quite different (Schusterman and Barrett, 1973). Also, the pigeon, and probably many other birds, are near-sighted for stimuli located directly forward and viewed binocularly, whereas they are far-sighted for stimuli located laterally and viewed monocularly (Catania, 1964).

A newer approach to the eye's spatial resolution is in the measurement of contrast-detection as a function of *spatial frequency*. A grating of stripes somewhat like those used in some acuity studies is shown to a subject whose response may be recorded behaviorally or electrophysiologically. The stripes, however, are not simple light and dark alternations; instead, the intensity varies continuously according to a sinusoidal function, luminance being highest in the middle of the light portion and lowest in the middle of the dark portion of the cycle. Even simple sinusoidal functions, however, look like fuzzy dark and light stripes instead of continuously changing intensity because the visual system is tuned to certain spatial frequencies. Spatial frequency perception is studied by varying the frequency of simple displays, combining frequencies (especially harmonics), varying the amplitude of luminance, and so on.

A surprising result has emerged from the study of *contrast-sensitivity* in spatial frequencies. Obviously, the ability to see the pattern depends upon the contrast-threshold (and hence its inverse, contrast-sensitivity). It turns out that the contrast-sensitivity depends upon the spatial frequency: sensitivity is maximum at some particular frequency and declines with increasing or decreasing frequency (*fig 5-15*, next page). To appreciate this phenomenon, consider a striped pattern with relatively low contrast. At an optimum distance, the striped pattern is detectable, but at a greater distance the spatial frequency is higher (more stripes per visual angle), so sensitivity declines and the striped pattern becomes invisible. Similarly, at close distances (lower spatial frequency with fewer stripes per visual angle), the stripes also disappear because the contrast falls below threshold. The phenomenon is so counter-intuitive that many persons find it hard to believe even when shown the proper stimuli. This phenomenon, plus species-differences in the optimum spatial frequency (*fig 5-15*) have important potential implications for

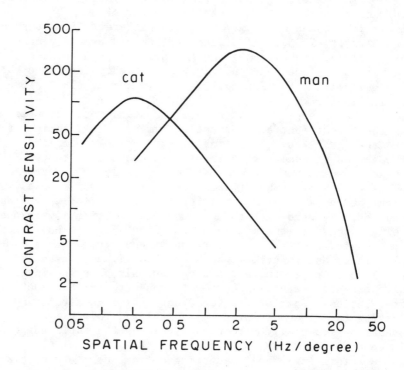

Fig 5-15. Contrast-sensitivity curves for two species (after Campbell and Maffei, 1974).

animal signals *(chs 7* and *8)*, as well as being an experimental tool for studying resolving phenomena of the eye.

As one might expect, a mosiac of small dots that are below the resolving power of the eye fuse perceptually. Black and white dots, such as used in photographs of newspapers, fuse into shades of gray. Colored dots fuse into intermediate colors including achromatic colors, which is the perceptual principle used in grain color films and in color television. The mixing of brightnesses or colors in such spatial arrays is appropriately called *spatial-mosaic fusion.*

---

pattern-perception

There is no unified explanation of how visual systems
recognize even simple two-dimensional patterns. Traditional associationist psychology held that the recognition
of even simple diagrams as distinctive wholes was slowly
acquired through learning, whereas ethologists and Gestalt
psychologists tended to consider the pattern-perception of
at least key social signals to be a property of neural
organization largely uninfluenced by learning (see Hailman,
1970 for a discussion). In recent years, cognitive psychologists have begun to favor the latter view, finding
that the human eye may see contours of shapes that are
not actually present in the visual stimulus (*e.g.*, Lawson
and Gulick, 1967; Gregory, 1972; Coren, 1972). For example, in *fig 5-16* one sees a white square that is not,
in fact, present. Such illusions do not depend on regular geometric shape of contours perceived in all cases,
but perception of shape does appear to involve "rules"

*Fig 5-16. A visual illusion, in which the eye sees a
white square.*

that are being systematically articulated. Perception of
two-dimensional pattern is not a simple process of learning to sort matricies of dark and light areas into categories that are subsequently recognized; rather, the visual system is predisposed toward processing sensory data
according to rather complex organizational principles.

Perhaps the most significant advance in the neuro-
physiological bases of pattern-perception is developing
from research on receptive fields of individual neurons
in the visual system of vertebrates. Each visual neuron
responds to key stimuli in only part of the total visual
field of the eye, the neuron's field being known as its
receptive field. Kuffler (1953) found that receptive
fields of the cat's visual neurons were not homogeneously
sensitive to light: some areas responded best to an increase in intensity and others to a decrease; some best
to light in the center, others to light in the periphery.
This breakthrough led to many investigations of receptive
field geometries and neural mechanisms in a variety of
other animals, including rabbits, turtles, salamanders,
pigeons, frogs, *etc*.

The basic organization of the vertebrate retina, although differing in details among species, may be summarized from studies of the mudpuppy (*e.g.*, Dowling and Werblin, 1969; Werblin and Dowling, 1969; Werblin, 1971, 1972).
The most direct path of neural data goes from the photoreceptors (rods and cones) via bipolar cells to ganglion
cells, whose axons make up the optic nerve and send data
from the eye to the brain. Therefore, a light stimulus
to a photoreceptor activates the ganglion cell(s) directly
in line with it in the retina. Light peripheral to this
central axis activates other photoreceptors, which are
connected not only to bipolar cells, but also to horizontal
cells that carry data laterally in the retina and may inhibit the photoreceptors or bipolar cells of the central
axis. Therefore, light stimuli directly in line with a
ganglion cell may excite it whereas light peripheral to
the axis may suppress its activity. A deeper layer of
amacrine cells carries even more peripheral sensory data
to the central ganglion cell. This basic organization of
the retina, which underlies the phenomenon of lateral inhibition mentioned earlier in this chapter, controls the
retina's adaptational state (also see above) and its relative general sensitivity to light.

If one records electrically from the ganglion cell, one expects to find a central area of the receptive field where light activates the cell and a peripheral area where light suppresses activity, and just this pattern is found in the cat *(fig 5-17a)*. Cessation of peripheral light is

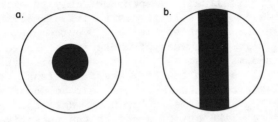

Fig 5-17. Two schematized receptive fields. A given neuron is stimulated by light in a restricted part of the animal's total visual field: the cell's receptive field. Certain cells respond maximally to a dark center with light surround (left), others with the reverse polarity. Higher in the pathway from receptors to brain other cells may respond maximally to dark bars in the receptive field (right), and different cells may have different preferences for orientation of the bar.

followed by an after-discharge of activity in the ganglion cell. Not only these "on-center, off-periphery" cell geometries, but also the reverse polarities are found in the cat. Not all vertebrate ganglion cells have such simple bullseye-like receptive field, however. In the frog, there are five classes of complexly organized types, such as the one that responds only to dark convex shapes that move through the ganglion cell's receptive field (Maturana *et al.*, 1960). In the pigeon's retina some cells are specialized to detect edges in particular orientation or movement in particular directions (Maturana and Frenk, 1963).

In a classic series of investigations, Hubel and Wiesel (*e.g.*, 1959, 1961, 1962, 1963, 1965) explored the receptive fields of cells more central in the visual pathways. After processing in the lateral geniculate, sensory data are sent to the visual cortex, where "simple"

cortical cells respond to particular orientations of slits
of light (or dark) in their receptive fields *(fig 5-17b)*.
Cells within one column of the cortex respond best to
slits in a particular orientation, those in other columns
to different orientations. Receptive fields at this level
in the cat are much larger than and apparently an integration of the bullseye-like fields aligned in the retina.
"Complex" cortical cells respond not only to orientation
of a light slit, but selectively to its direction of movement within the receptive field. "Hypercomplex" cells respond to a slit regardless of where it is placed within
the field and may also show preferences for particular
lengths of slits. In the frog, central visual cells have
such complex properties that they must be described in
nearly ethological terms: a "newness" detector responds
when an insect-like stimulus enters the receptive field
and a "sameness" unit responds as long as the stimulus
remains, even if it ceases to move--yet responds more
vigorously when it moves again (Lettvin *et al.*, 1961).

From such studies of the receptive fields of individual cells in the vertebrate visual systems one may already reach several general conclusions. (1) Visual systems are highly organized to extract information about
two-dimension patterns of light, implying that some patterns will be easier to detect than others. (2) As one
moves centrally within the visual system, cells respond
to more specific kinds of stimuli, and at one level it is
possible to have different cells responding preferentially
to different kinds of key stimuli. (3) The organization
differs in different species, so without specific studies
it is difficult to suggest what patterns will be most effectively detected. In visual systems used primarily
for specific visual tasks--such as detection of predator
and prey in frogs or coordination of flight in pigeons--
the organization of receptive fields appears to be more
specialized, although too few species have been studied
to yield final conclusions.

The commonest receptive-field geometry is the bulls-
eye type, which seems almost an inevitable consequence of
lateral inhibition working through the anatomical substrate of the retina. Bullseye-like receptive fields are
known primarily from ganglion cells, but also from cells
at higher centers, in a variety of animals including spider
monkey (Hubel and Wiesel, 1960), rhesus macaque (Wiesel
and Hubel, 1966), domestic rabbit (H.B. Barlow *et al.*,

1964), laboratory rat (Brown and Rojas, 1965), Mexican ground squirrel (Michael, 1968) and goldfish (Jacobson and Gaze, 1964). Perhaps because of this widespread organizational principle in receptive fields, bullseye-like stimuli are readily recognized by animals (see Wickler, 1968: 64-70). So far as I am aware, there has been only one attempt to connect receptive-field geometries with the characteristics of optical social signals (Hailman, 1971; see also figure 15 in Hailman, 1977a for a summary), but the topic may be a fruitful one for future research.

Whatever the exact organization of sensory inputs, the visual systems of invertebrate and vertebrate animals alike enhance the boundaries between light and dark in spatial patterns by the process of lateral inhibition, mentioned previously. An important consequence of lateral inhibition has been recognized for a long time in human vision (see Minnaert, 1954: 105-106). The boundary between light and dark is not only enhanced, but also shifted toward the darker side, a phenomenon confusingly called *irradiation* (to be distinguished from irradiance, *ch 3*). Most persons are familiar with irradiation, even though they may not have realized it. When dark telephone lines cross against a bright sky, the point of intersection disappears, and there seems to be a gap in the wires. Or, when the sun rises (or sets) over water, the horizon seems to dip in front of the sun. It is possible that this principle is utilized in some optical signals *(ch 8)*.

depth-perception

Man and other animals often have good visual judgment concerning the distance of some perceived object. Psychologists have admirably elucidated the multiplicity of visual clues used in such *depth perception,* only some of which depend upon possessing two eyes. Almost a dozen sources of depth information have been described.

One class of cues concerns properties of the environmental scene that can be captured in still photographs. These are the cues that provide the illusion of depth in pictorial representations on a flat surface. *Perspective* or *convergence* is the cue provided by parallel lines meeting at the horizon, such as railroad tracks. A related cue is the *known size* of an object, whose image becomes smaller at greater distances. There is a case that falls between

perspective and known-size cues: repetitive units that
become smaller with increasing distance, such as telephone
poles. Imaginary lines connecting their bases and apexes
meet at the horizon, providing a perspective cue that may
operate in conjunction with the learned size of telephone
poles. Another type of clue, which may be considered a
subset of convergence, is *texture gradient*. Because of
perspective, the units of texture--such as the cobblestones
of a street--become smaller in the distance.

A third major photographable cue is *superposition*, in
which nearer objects may partially obscure the view of more
distant objects behind them. A fourth, and rather subtle,
cue is *elevation*. Objects higher in a scene, particularly
if they extend above the horizon, appear farther away than
lower objects. This principle follows logically from perspective, but perceptually it appears to be distinct.

Two photographable cues depend upon environmental conditions: *brightness* and *distinctness*. Closer objects tend
to look brighter than farther objects because of absorption
and scattering by the medium between object and viewer *(ch
3)*. This cue is particularly useful in turbid waters, as
every SCUBA diver learns. Similarly, nearby objects appear
more distinct because their images are less blurred by the
medium than are those of farther objects.

To these six depth cues that can be captured photo-
graphically may be added several others that rely on mechanisms of perception. *Accommodation,* the focusing of an
image onto the retina, provides a seventh cue. In order to
accommodate, the dioptics of the eye must be altered (such
as changing the shape of the lens) for different distances.
In man, a distant object demands a thin lens and a nearby
object a thick lens for proper accommodation. If the observer can internally sense the thickness of her lens *(e.g.,*
by spindles in the muscle fibers that alter shape), this
feedback provides a depth cue.

Some cues derive from viewing the world from different points in space. In *motion parallax*, nearby objects
such as fence posts along a road travel rapidly to the rear
of one's visual field, whereas distant objects such as hills
move backward more slowly. (Very distant objects such as
the moon appear to be stationary.) A simple form of motion
parallax is *displacement parallax*, in which a one-eyed animal can look at a scene from one place and then move its
head sideways to look from another spot: closer objects
change their spatial relationships with one another more

drastically than do farther objects, which tend to stay in
the same place within the visual field.

Animals with two eyes can compare both views at once,
the differing images on the two retinas being called *binocular disparity*. Furthermore, in order to accommodate
nearby objects, the two eyes must be rotated inward, and
this cue of *ocular convergence* is apparently perceived by
feedback from sensory cells in the muscles that control
eye-movements.

There is an illusion of depth due to color used by
some painters (presumably consciously). This chapter a-
voids dioptic considerations of eyes because the structure
of animal eyes varies so greatly, but the color-illusion
of depth demands an excursion into structure. The principal refracting surface in the vertebrate eye is the outer
surface, called the *cornea (fig 5-18a*, next page). The
lens within the eye is either moved backward and forward
like a camera lens (*e.g.*, in birds) or its shape is changed
(*e.g.*, in man) to effect the fine adjustments in accommodation (focusing). Because the index of refraction varies
with the frequency (*eq 3.8*), different colors are brought
to focus at different distances behind the lens (*fig 5-18b*).
This is the phenomenon of chromatic aberration mentioned
in *ch 3*.

The human eye accommodates through changes in the
shape of the lens, so that it must be made thinner to focus violet rays on the retina (*fig 5-18c*) and thicker for
red rays (*fig 5-18d*). However, a thin lens also brings
distant objects into focus (*fig 5-18e*) and a thick lens accommodates nearby objects (*fig 5-18f*). As noted above, the
eye can sense these changes in accommodation for distance
through internal receptor cells in the accommodation mechanism, and use such sensory data to judge distance. Therefore, accomodating violet surfaces provides internal data
signaling distant objects and accommodating red surfaces
sends data signaling nearby objects. I delight in viewing
my reproduction of Paul Klee's "Around the Fish," a painting of red fish on a blue platter in which the fish appear
to stand out from the platter. This illusion, although
often subtle, may be used in the design of some optical
signals (*ch 8*).

perceptual constancies

A remarkable aspect of vision is that certain per-

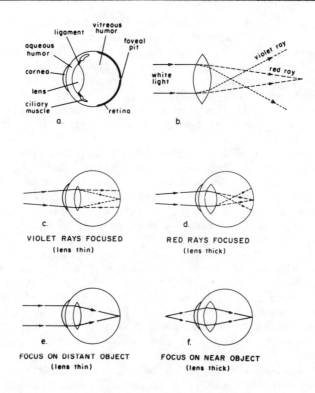

Fig 5-18. Depth illusion based on color due to chromatic aberration in the eye. The cornea is the principal re-fracting surface of the human eye (top left), but fine changes in accommodation (focus) are effected by changes in the shape of the lens. The lens, however, shows chro-matic aberration (top right), so that violet rays require a lens-shape (middle left) like that used to focus on distant objects (bottom left), and red rays require a shape (middle right) used to focus on nearby objects (bot-tom right). For this reason, a red spot may seem closer to the observer than its blue or violet background when the colors are actually on a plane. (Angles of rays and shapes of lenses exaggerated greatly in diagrams.)

ceptions occur without an immediate sensory basis, and must depend upon simultaneous contrasts of some complexity or upon past experience, or both. One perceives a basketball as spherical, even when there are no evident shadows to provide clues as to its three-dimensional shape. This *shape constancy* is due in part to having viewed basketballs from every direction, and in so doing having found that they always have a circular pattern. Apparently, much of our perceptual identification of three-dimensional shape depends upon such experience in viewing objects from many angles, and forming a mental picture of the shape.

Constancy of brightness is very persistent. A white triangle upon a gray background still appears like a white triangle on a gray background when viewed in extremely dim light--even though the gray in bright light may be absolutely brighter than the white in dim light. Part of this constancy may be explained by adaptational shifts of the eye, but after these are taken into account the constancy still remains, and therefore must be due to the simultaneous contrast between the two surfaces. For this reason, the apparent surface brightness of an object is not directly proportional to its photometric luminance.

Similarly, objects appear to have a relatively *constant color* under differing conditions of illumination, although homogeneously colored surfaces viewed in isolation lose such color constancy. SCUBA divers identify color patterns of fishes at various depths where the actual spectral composition of the stimuli varies markedly. This phenomenon of spatial induction of color sensations has proved a continuing problem in the generation of a comprehensive theory of color vision.

Most of the research on perceptual constancies relates to human vision or that of a few domesticated laboratory animals. There is very little known about the roles of simultaneous contrast and past experience in animal vision, yet various kinds of perceptual constancies might be important to understanding the design of optical signals. For example, certain signal shapes might be designed by evolution to be recognizable from various angles, or color patterns might be designed to be recognizable under varying conditions of illumination. Such possibilities are deserving of experimental attention.

Overview

Unlike the channel *(ch 3)* and the sender *(ch 4)*, the receiver imposes many important limitations upon the encoding of information in optical signals. Visual pigments, which absorb photons entering the eye, are more sensitive to some frequencies than others and have a limited dynamic range of photon-fluxes over which they can generate differential neural signals. The spectral and intensity ranges of the eye may be extended by having populations of different receptors that operate best in low or high light levels, or which have sensitivity curves located in different parts of the spectrum. At least three of the latter such receptor types are needed for reliable color vision, but even with three types color-equivocation may occur. Perception is built upon spatiotemporal arrays of photon-fluxes, but its characteristics are many and complicated. There are both temporal and spatial resolving limits that constrain the receiver's ability to accept information encoded temporally and spatially in light. Many perceptual phenomena, particularly those involved with illusions and constancies, suggest that perception is a complex, active process that depends upon events in the entire visual field as well as patterns of expectation in the visual processing system itself, some of which are established through prior visual experience. This chapter could do no more than point out some of the aspects of animal sensation and perception that must be taken into account if one is to understand the design-characteristics of optical signals. Only the most evident and well-known aspects of vision are mentioned; many others may well play a role in understanding optical design of signals.

Recommended Reading and Reference

This chapter differs from others in that it reviews only selected topics from its potential subject matter—those I felt necessary for understanding later chapters. Therefore, the *Appendix* presents an unusually long list of relevant articles from *Scientific American* through which the reader can expand upon the subjects of this chapter without delving into highly technical literature. A few technical volumes, though, are worth mentioning. *The Visual Pigments* by Dartnall (1957), although outdated, is still a clear introduction to pigment chemistry. Rodieck's (1973) *Vertebrate Retina* provides excellent coverage of its subject. The vision volumes of the *Handbook of Sensory Physiology* (*e.g.*, Dartnall, 1972; Fuortes, 1972) cover virtually all the physiological aspects of vision mentioned

in this chapter, as well as much other material. The most
authoritative reference on photometry and colorimetry is
still LeGrand's (1957) classic *Light, Color and Vision*,
but a book by one of LeGrand's translators covers much of
the same ground and is available in a Dover reprint (Walsh,
1958). An excellent reference that ranges from molecular
aspects of vision to the complexities of visual perception
is Pirenne's (1967) *Vision and the Eye*. There are, of
course, many other fine sources of information about vision
and visual perception.

We can really learn the truth about the e-
volution of signals best from the liars.
                              --Wickler (1968: 234)

This chapter treats the visual principles by which
an animal deceives an observer. Often, visual deception
is partitioned into *concealment*, in which an animal goes
undetected in its environment (Cott, 1957), and *mimicry*,
in which it goes unrecognized for what it is (Wickler,
1968). However, there is no logically sharp distinction
between the two, nor do these topics include all deception. In some forms of deception the animal is detected
and recognized for what it is, but the observer is deceived about its probable subsequent actions. Other forms
of deception involve only parts of the animal, as in concealment of the eye. *Deception* is here taken to include
any transmission of *"mis*information" (see *ch 2*).

Already there exists confusion over terminology with
which transmission of "misinformation" is designated.
Wickler's quote above suggests that all deception could
be called *lying*, but Sebeok (1975) recommends restriction
of that term to "mendacity in language" and prefers *pre-*
varication as the generic term for deception "involving
any sort of sign." However, prevarication also implies
linguistic behavior and furthermore connotes evasion of
truth by quibbling or dodging. Therefore, I use *deception*
as the generic term, *feigning* as the specific term for
behavioral deception and *lying* as the specific term for
linguistic deception. I suggest that *prevarication* be

-157-

reserved for cases of evasiveness in linguistic deception, and will have no further use for it in this volume.

Escape from predation is a principal selective factor producing concealing deception, but there are many others as well. The predator benefits from being undetected or unrecognized by its prey, and animals may also benefit from deceiving cleptoparasites and competitors of the same or different species (Hailman, 1963). A cleptoparasite steals directly from its host, as in eagles stealing fish from ospreys or ducks (Grubb, 1971), whereas a competitor steals indirectly by utilizing the same resource. The contended resource is often food, but may be nesting sites, potential mates and so on. In this chapter I am not concerned specifically with the selection pressures favoring deception, but rather with the visual principles upon which the deception is based and their implications for animal behavior and morphology.

To provide a common vocabulary, it is useful to point out a few terms used frequently after this point. Adaptations that render an animal difficult to detect have been termed concealment, protective coloration, crypticity, camouflage, *etc*. *Protective coloration* omits behavioral and structural aspects of deception and carries the restricted connotation of predator-prey interactions. *Crypticity* has the connotation of mysteriousness and *camouflage* has a military connotation, so I prefer *concealment*. The use of *mimicry* and *imitation* are discussed below (see section on *imitation*).

An animal effects deception by behavior or morphology, the latter being divided into physical structure (*e.g.*, shape) and surface coloration. By *coloration* I mean all properties of the surface that affect reflection (or transmission) of light, including not only spectral reflectivity, but also diffuse-*vs*-specular reflectance, index of refraction, *etc*. *Contrast* is used in the broad sense of a visual difference in spectral radiance between two surfaces (*ch 3*); it includes not only differences in luminance ("brightness-contrast," *ch 5*) but also differences in spectral composition ("color-contrast" involving the sensations of hue and saturation: *table 5-I*, p. 138).

It might seem that deception is tangential to the subject of visual communication, but this is not so for at least four important reasons. (a) If one is to understand the design of optical signals, it is necessary to consider all factors that affect animal coloration and associated behavior; hence the previous chapter and this one are particularly crucial. (b) Visual deception *is* optical communication, albeit it commonly *inter*specific communication of *mis*information. Moynihan (1975) suggested that adaptational complexes designed to conceal the animal be called *anti-displays*, in distinction to traditional displays that render the animal conspicuous. (c) By articulating the principles of concealment, one may identify their inverse principles of conspicuousness used in social displays (see next chapter). Finally, (d) it is possible that the *same* behavior or morphology may be used to communicate "misinformation" to one receiver (such as a predator) and "correct" information to another (such as a social companion). Discussion of how animals might simultaneously be concealing and conspicuous is deferred to the next chapter.

There appear to be at least four natural categories of visual deception. In the first, (1) an animal diminishes the probability of detection by suppressing the effects of shadows it casts on its own body and on the substrate. Or, (2) an animal diminishes detection by suppressing the clues by which its outline can be detected. These two are classical concerns of concealment. In the third category, (3) an animal diminishes either detection or correct recognition by an observer through imitating the visual characteristics of something else, a subject that bridges concealment and mimicry as traditionally partitioned. (4) The last category diminishes detection or recognition by creating visual ambiguities and perhaps employing other factors as well. The chapter considers each of these four strategies in turn, and the known mechanisms by which they are effected.

Finally, it should be noted that throughout most of the literature on deception, the mechanisms employed by animals have really been judged against human vision. Recently, Pietrewicz and Kamil (1977) have reported operant conditioning experiments in which blue jays were trained to detect underwing moths in colored slides, and hence relative concealment could be judged objectively according to the vision of a relevant observer, in this case a predator. More experiments like these are needed.

Suppression of Shadow-contrasts

An object in sunlight casts a shadow on its own lower surface as well as upon the substrate *(figs 3-11 and
3-12)* and hence an observer can detect the object either
by the luminance-contrast between its upper and lower surfaces or by the luminance-contrast between its substrate-
shadow and the adjacent substrate. The obvious strategy
for suppressing these contrasts is to eliminate the shadows or diminish their size or contrast.

shadow-minimization

There are at least four ways to minimize the shadow
cast upon the substrate, and the first three minimize shadow on the ventrum as well. (1) The animal may avoid loci
of strongly directional light, for example, by remaining
in the shadow of other objects or by frequenting shadow-
less environments such as sufficiently deep water where
scattering eliminates directionality *(ch 3)*. Thus, deep
woodland birds and many fish cast virtually no detectable
shadows because of *directed-light avoidance*.

2) Laterally compressed animals cast very small shadows on the substrate and almost none on themselves, and
hence also have little problem being detected by shadow-
contrast. Many fishes are so shaped, as are many insects;
other insects at rest fold their wings vertically above
their bodies *(e.g., butterflies and damselflies)* to create
lateral body-compression.

3) By remaining closely appressed to the substrate
an animal eliminates substrate-shadow, and if flattened
dorso-ventrally eliminates or greatly minimizes shadow
on its ventrum. A typical example is the moth that ap-
presses itself closely to the bark of a tree. Flounders
(Bothidae and Soleidae) are an extreme case, in which the
compression may be due to other factors as well. Of course,
lateral compression has the effect of dorso-ventral compression if the animal lies on its side, as do many flatfishes. My own observations on shorebirds on the open
beach and mudflats suggest that when resting they often
lie on the substrate instead of standing like many other
birds, possibly an example of this tactic of *substrate-
appression*.

Finally, (4) an animal can eliminate substrate sha-

dow by simply remaining far from a substrate *(fig 3-12,*
p. 77). However, to eliminate its umbra totally an animal must be more than 100 body-lengths from the substrate
(eq 3.11), although its shadow will decrease in size as
a linear function of the animal's distance from the sub-
state below *(eq 3.12)*. Insects can fly high enough to a-
void casting shadows on the earth's surface, but ordinary
(non-migratory) flight of most birds is below 100 body-
lengths in altitude. It seems likely that because of
scattering *(ch 3)* the umbral shadows cast under water are
less distinct than in air, so that fishes and other aquatic animals may be able to remain well within 100 body-
lengths of the bottom without casting definite body shadows. Furthermore, surface wave-action is liable to disrupt shadows in shallow water, so that shadows in aquatic
environments will probably rarely be important. In general, then, *substrate-avoidance* may be a limited behavioral
mechanism for avoiding tell-tale shadows.

In sum, substrate-shadows may be a relatively minor
problem among animals. The shadows are most important
when the animal is large in body-size relative to the ir-
regularities of the substrate upon which its shadow falls,
and here only in direct sunlight. In this cases, substrate-
appression may be expected to be most common, and when the
animal's dorsal coloration matches that of the substrate
such behavior also makes its profile less detectable (see
below).

counter-shading

A.H. Thayer seems to have originally proposed about
1896 that contrast between dorsum and ventrum due to the
animal's own shadow could be diminished by *counter-shading*,
in which the dorsum has a lower reflectivity than the ventrum. The idea was developed by his grandson, G.H. Thayer
(1918), discussed extensively by Cott (1957) and shown
conclusively by De Ruiter (1958) and others to promote
concealment of caterpillars from their predators.

A few considerations suggest where counter-shading
should be effective: (i) the animal's body must be shaped
so as to cast a shadow on itself, and (ii) its illumination must come habitually from above. These conditions
are so readily met by terrestrial animals that "normal"
coloration incorporates counter-shading in many insects,

amphibians, reptiles, birds and mammals. Indeed, it is
the exceptions that arouse curiosity, such as the conspicuous reversed counter-shading of the breeding male bobolink, which is tan and white above but jet black below
(see *ch 7*). Caterpillars that frequent the undersides of
leaves and twigs are also reverse-countershaded, but their
upside-down posture renders them truly counter-shaded in
the environment. A fascinating example from fishes is the
upside-down catfish, which true to its name swims upside-
down, putting its dark belly upward (Sterba, 1962: 409 and
plate 95).

For cylindrically shaped bodies, one expects the
counter-shading pattern to be darker above the lateral mid-
line, but in other cases the pattern may be different. If
the animal is roughly rectangular in cross-section, only
its belly is in shadow, and counter-shading is similarly
patterned. For example, the white-tailed deer is brown to
gray on its back and deep vertical sides, being white only
on the belly. The exact pattern of counter-shading must
therefore be examined in each case with regard to the animal's shape and the shadow it casts on its own parts.

The strength of the contrast between dorsum and ventrum in counter-shading should be a function of the degree
to which illumination is directed from above. For animals
living in open sunlight, one expects and finds strong contrast, but in animals adapted to deep woodland conditions
or in open waters having much scattered light, the contrast
may be more subdued. We expect counter-shaded fish only
in relatively shallow water, and the coney is an excellent
example. This species has been collected in many color
phases, some of which are uniformly yellow to brownish red,
but one of which is strikingly counter-shaded. Chaplin
(1972: 14) notes that this latter is a "common shallow water form" and Böhlke and Chaplin (1968: 264) note that
"this phase has been termed the 'excitement phase' and
thought to occur when the fish is alarmed or is stimulated
by the presence of food, but we have observed it as a long-
term phase shown by fish cruising over the reefs." *Figure
6-1* compares counter-shading with other mechanisms for
suppressing contrasts due to shadows.

A few notes might be added to the subject of counter-
shading. First, other factors--such as protection from
UV-radiation *(ch 4)*--predict the same pattern of coloration,
so that the counter-shading explanation should always be
treated only as one initial hypothesis in any specific case.

SUPPRESSION OF SHADOW-CONTRASTS

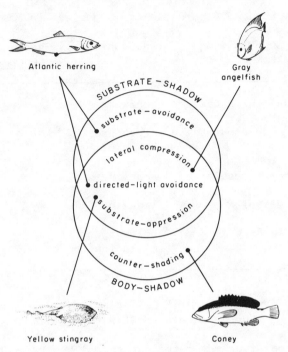

Atlantic herring

Gray angelfish

SUBSTRATE-SHADOW

substrate-avoidance

lateral compression

directed-light avoidance

substrate-oppression

counter-shading

BODY-SHADOW

Yellow stingray

Coney

Fig 6-1. Suppression of shadow-contrasts exemplified by various fishes. Substrate-avoidance suppresses only substrate shadows and counter-shading suppresses only body-shadows, but the other three mechanisms suppress both kinds of shadows. Some animals may exhibit more than one mechanism simultaneously.

Second, reversed counter-shading leads to visual conspicuousness, and therefore may be used in social communication as a signal *(ch 7)*. Finally, it is interesting to note a special kind of "counter-shading." Some bioluminescent fishes utilize ventral photophores to make the lower part of the body about as bright as the upper part (Hastings, 1971). This is only one of many uses to which bioluminescence is put by animals (see Lloyd, 1977). The types of suppression of shadow-contrasts used by animals are summarized in *table 6-I* on the following page.

Table 6-I

Types of Suppression of Shadow-contrasts

Type of Suppression	Adaptation(s)	Shadow-contrast Suppressed	
		substrate	body
directed-light avoidance	behavior	X	X
lateral body-compression	structure behavior	X	X
substrate-appression	behavior*	X	X
substrate-avoidance	behavior	X	–
counter-shading	coloration bioluminescence	–	X

Enhanced by dorso-ventral structural compression

Suppression of Outline-contrasts

An observer may detect an animal by its contrast in spectral radiance (hue, saturation or brightness sensations) relative to the background. The contrast is greatest at the periphery of the animal's image due to lateral inhibition *(ch 5)*, so the visual clue is the animal's shape or outline. This tell-tale outline may be suppressed by at least three optical strategies: distracting attention away from it (disruptive pattern), eliminating it (transparency) or diminishing it by resemblance in spectral radiance to the background (matching coloration).

disruptive coloration

The outline of an animal may be disrupted by presenting a different outline that is visually more compelling. If the alternative outline resembles some recognizable shape, the deception works by confusing the observer's recognition (see *imitation*, below). This section concerns deception that simply disrupts the tell-tale outline and

hence diminishes the probability of the observer's detecting the animal against its background.

The visual principles of such *disruptive* (or ruptive)
coloration appear to be unspecified, so it is worth attempting to identify some of them. (1) It seems likely
that the contrast between surface patches of the object
must be at least as great as the contrast between a patch
and the background. Restricting the case to a two-color,
achromatic animal, the ideal relation seems to be

$$R_1 \geq R_b \geq R_2 , \qquad\qquad (6.1)$$

where R_1 and R_2 are the reflectances of the patches and
R_b is the reflectance of the background. Any other relation will render the contrast between at least one patch
and the background stronger than the contrast between the
two patches. Therefore, ideal disruptive coloration might
be black and white, or at least dark brown and light yellow.

2) Unless the pattern deceives by resembling something else (see *imitation*, below), the patches should be
randomly shaped, or at least irregular with respect to the
animal's true shape.

3) The patches must be arranged such that light and
dark alternate on the periphery of the animal as seen by
the observer. Otherwise a light edge or dark edge outlines
the animal and makes it *more* conspicuous (Cott, 1957).

4) It seems likely that if the patches are too small
the animal looks patterned and may even look uniformly
colored at a distance because of spatial mosaic fusion
(p. 144). Similarly, if the patches are too large, the
animal's real shape may be evident simply because the a-
mount of deceptive contrast is so limited. Therefore, there
should be an ideal size of disruptive patches, relative
both to the distance at which the object is viewed and its
own size. T. Johnston *(pers. comm.)* suggests further that
the ideal disruptive pattern should match the "graininess"
of the background against which it is seen.

Many patterns cited as disruptive also involve other
principles of visual deception. For example, military
camouflage of tanks and gun-implacements often involve
matching vegetation color (see *matching coloration*, below)
as well as disrupting tell-tale shapes. During World War
II, warships were painted in disruptive patterns. These
have since largely been abandoned, perhaps because the

patterns were not effective deception, or more likely because with the advent of radar visual detection of ships
is no longer a consideration.

   Ostensible disruptive coloration of animal may often
(like military camouflage) involve other principles, but
convincing examples of disruption are common (Cott, 1957).
Most, like the breast-band on many small plovers (Charadrius), are black and white, conforming to the first principle deduced above. The orange dog caterpillar of the
giant swallowtail butterfly is dark brown and pale yellow,
the patches being quite irregular in shape, thus conforming to the second principle. Huxley (1958) suggested that
the killdeer, which is an exceptionally large plover, possesses *two* black breast bands because of its size (third
principle). The breast-bands of plovers are not, however,
irregularly shaped, nor are the markings of many boldly
patterned fishes, so one suspects in such cases additional principles of deception are being combined with disruptive coloration. G.W. Barlow (1967) has suggested that
the middorsal stripe on the head of certain fishes is disruptive coloration: the "split-head" pattern.

transparency

   Perhaps the most straightforward solution to suppressing an animal's outline is to eliminate it by making the
animal transparent so the observer sees the background
through the animal. Certain optical and morphological considerations suggest where animal transparency is most likely
to be found. It is not sufficient for an animal simply to
transmit light: even highly transparent substances such as
glass refract *(ch 3)* because they are optically denser than
the medium. The distortion due to refraction of an animal
may be minimized in at least two different ways: either the
animal evolves an index of refraction close to that of its
medium, or else it remains very near a background so that
visual distortion is minimized. Both strategies appear to
be used, at least sparingly, for visual deception.

   All animals, because they are constructed of macro-
molecules with high molecular weight, probably have relatively large indices of refraction. Highly transmitting
organic substance such as oils and fats have refractive
indices generally between 1.4 and 1.5. Therefore, it seems
almost impossible for an animal living in air (having a

refractive index near unity) not to refract light. However, water (n = 1.32 to 1.33, depending upon temperature)
provides a suitable medium, providing the animal has a
high enough water content to bring its own refractive index reasonably close to that of the medium. It is not
surprising, then, that most transparent animals are aquatic, familiar examples being certain coelenterates, ctenophores and many smaller animals that drift in the ocean,
as well as some adult fishes and fry, and a few other
freshwater animals. Of animals living in air, the wings
of dragonflies approach transparency, but similar examples
are relatively few.

There is a second optical principle of transparency
that may be utilized by certain aquatic animals. As pointed out by Ruechardt (1958: 30-31), an object that has a
different index of refraction from its medium can still be
invisible if illuminated homogeneously. He outlines a
simple experiment whereby a transparent glass rod may be
rendered invisible in air--an experiment that astounds
even those familiar with optical principles. Place the
rod inside an inverted cone painted highly reflecting white
inside, and illuminate from above; by peeking through a
pinhole in the cone one looks at the rod nearly homogeneously illuminated and finds it difficult to detect. Spatially homogeneous illumination does not occur in terrestrial environments, but in deep water illumination approaches
this ideal *(ch 3)* so that transparent animals with indices
of refraction different from that of water may still become nearly invisible.

Transparent animals living in air must adopt the strategy of remaining very near some background, since the highly refractive animal would be easily detected in free
space. J. Baylis and E.H. Burtt *(pers. comm.)* have pointed
out that the wings of certain dragonflies (Suborder Anis-
optera) and clear-winged moths (Aegeriidae)--which are held
horizontally while at rest--are sufficiently transparent
to be visually undetected. Wings of many species have dark
markings that are thus visually disconnected from the resting animal, but become visually conspicuous when they flicker in flight. It should be pointed out, however, that wings
of the related damselflies (Suborder Zygoptera) are held
dorsally over the resting animal despite being similarly
transparent (but see *shadow-minimization*, above).

It is also useful to note that animals may be almost
entirely transparent, as ctenophores, or have only certain

transparent parts, as in the fins of certain fishes. By
being partially transparent an animal may alter its shape
as perceived by another, and therefore may go unrecognized.
For example, a leaf fish with transparent fins does not
have the visible shape of a fish, and hence may go unrecognized by predators and prey alike. Finally, transparency is not always due to selection for visual deception.
For example, certain bathypelagic fishes are transparent,
but live in environments where light levels are effectively zero. It seems reasonable to assume that the metabolic
cost of producing unnecessary pigment is sufficiently high
to be selected against in animals living in poor trophic
environments.

matching coloration

The third solution to suppressing the visual outline
of an animal is for its perceived coloration to match that
of its background. There are actually two ways of making
the spectral radiance of an animal match that of its background: by achromatic (neutral) reflectance and by spectral reflectance similar to that of the background's.
My notion of "spectral conformity" of achromatic reflection comes from noticing the widespread occurrence of
white ventral ground coloration in birds. Although light
ventral coloration is a factor in counter-shading (see a-
bove), it need not be white to be effective. White coloration, being a selectively neutral reflector, will tend to
take on the hue of ambient irradiance and under specified
conditions will also match the background hue. Two conditions must be met: (a) the reflectivity of the animal
must be neutral (achromatic) and relatively high, or else
the animal will be colored (chromatic reflectivity) or gray
to black (low reflectivity); and (b) the ambient irradiance
reflected from the animal must resemble in spectral distribution the ambient radiance emanating from the background.
Put differently, the second condition means that the light
must be colored by its having been reflected or scattered
from the background, or else filtered in such a way that
the filtered light has the same spectral distribution as
that emanating from the background.
One common example fulfilling both conditions is the
white underparts of many forest songbirds. The ambient illumination is greenish because of transmission by canopy

leaves, and light reflected from lower leaves is even greener in spectral distribution. (Not all substances transmit and reflect the same spectrum, but leaves do.) Therefore, the light striking the white breast will be greenish in spectral composition, and hence the breast will reflect the greenish emphasis. Because the reflecting leaves are receiving approximately this same irradiance, and then further absorbing, they will appear both darker than the white bird and of a more saturated hue. Therefore, the deceptive effect is subtle and probably useful only in conjunction with other principles of visual deception, such as counter-shading.

The same principle should apply to animals that frequent a homogeneous substrate in open sunlight, if they remain close to the substrate so that their undersides are irradiated primarily by reflected light. The white breasts of small sandpipers and plovers thus utilize such reflectance from sand and mud, whereas longer-legged relatives may stand too high off the ground to be illuminated primarily by substrate reflectance. (For example, the curlews, *Numenius* spp., tend to be quite dark below.)

It seems likely that white reflectivity is also a factor in the neutral reflectances of so many silvery colored fishes. Open water species, unless they are near the surface, will be irradiated largely by bluish light scattered by the water. By reflecting this light achromatically, the fish blends well into the background of scattered light.

An obvious way to blend into the background is to match it in spectral reflectance characteristics. Such *chromatic reflectance* may involve one homogeneous reflection spectrum (uniform chromatic reflectance) or a patterned reflective surface resembling the background (patterned chromatic matching).

Uniform chromatic reflectance occurs when a homogeneously reflecting surface has the same spectral reflectivity as its common background, so the advantage of such a deceptive strategy over achromatic reflectivity is that the match can be quite convincing visually, but the disadvantage is that the match works only against one background. Uniform chromatic reflectance is so widespread and frequently documented (*e.g.*, Cott, 1957; Wickler, 1968) that only a few examples need be pointed out. Many arctic and alpine animals that live habitually on or near snow are permanently white (*e.g.*, the mountain goat and snowy owl) whereas others are white in winter but turn a matching brown in

summer when the snow is gone (*e.g.*, the snowshoe hare and
white-tailed ptarmigan). Dice (1947) was the first to
show experimentally that predators find prey more readily
when the latter contrast in brightness with their substrate.
Perhaps the most elegant and well-documented uniform matching is the case of industrial melanism, in which the European peppered moth living in areas where carbon deposits
cover the environment have become melanistic (*e.g.*, Ford,
1964; Kettlewell, 1973). Uniform chromatic reflectance is
commonly utilized in conjunction with other deceptive principles, as well as dictating the entire body coloration.
Industrial melanism also occurs in North American moths
(Owen, 1961).

One study of matching uniform chromatic reflectance
impresses me as important in showing the way toward future studies. Johnson and Brush (1972) have studied the
polymorphic coloration of the Central American sooty-capped
bush tanager. First, the coloration itself is documented
objectively with reflection spectrophotometery *(ch 3)*.
Second, the C.I.E. chromaticity variables *(table 5-I)* are
used to describe the visual appearance of the birds, and
then used analytically to establish the pigmentational basis. The darkness of the tanagers is due to concentration
of the biochrome (lutein), and as the concentration increases the excitation purity increases linearly with it,
indicating a single-pigment basis for coloration. Geographic variation is then used to show that the dark gray-
green morphs are restricted to high volcanic mountains where
"the background of ash, vegetation, and fog provides an
environment of pervasive grayness." Employing the comparative method *(table 1-III)*, Johnson and Brush find many o-
ther avian species with gray plumage in this habitat, and
when the species range widely into other areas they are
most common in the gray environment. Finally, as a tanta-
lizing footnote, there is some evidence that the gray-green
phase of the tanager increased in abundance following the
major eruption of Volcan Irazu in 1963. This kind of careful study sets standards for future research on animal
coloration.

Some caution is necessary in identifying matches in
chromatic reflectance. We humans tend to classify everything according to our own sensory systems, but Eisner *et
al*. (1969) provide a clear example of problems with this.
Crab spiders, which frequent floral heads, may match the
flower coloration to our eyes and hence be concealing to

predators such as birds. However, when viewed with an ultraviolet analyzer, they may be quite conspicuous and hence
could be easily detected by their insect prey having good
UV-sensitivity *(ch 5)*.

A special kind of uniform chromatic reflectance seems
worthy of mention because it appears to be widespread.
Many insectivorous birds have dark lines or patches around
the eyes, which themselves have dark irises, so that the
tell-tale outline of the eye is not revealed. The eye is
in a black field in many woodpeckers, flycatchers, swallows,
corvids, chickadees and so on, although other explanations
for black about eyes are possible *(ch 4)* and no critical
studies distinguishing the hypotheses seem to exist (also
see discussion in Burtt, 1977).

When the substrate is patterned rather than being uniformly reflecting the animal may evolve *patterned chromatic reflectivity* to match it. Many ostensible cases may
involve disruptive principles rather than strict matching,
but there are convincing examples. Wickler (1968: 55, figure 6) shows a cogent photograph of the sole against a sand-
grain background, and many flatfishes (flounders, Bothidae;
soles, Soleidae; and tonguefishes, Cynoglossidae) resemble
or can change color to resemble their patterned substrates.
The light form of the peppered moth in non-industrial areas
is virtually invisible on the patterned bark of oaks (see
photographs on p. 91 in Bishop and Cook, 1975). G.W. Barlow
(1974: 27) points out that the surgeon fish *Acanthurus gut-
tatus*, which is found in the surf with many surface water-
bubble, is marked with numerous white spots. Sometimes variation within a species yields a correlation between habi-
tat-choice and concealing coloration. Schoener and Schoener
(1976) found that small female brown lizards perch on small
branches and have longitudinal stripes whereas larger females perch elsewhere and are unstriped.

In some cases, special *matching structures* may accompany matching patterns. For example, some small inshore
bottom fishes of California show blotchy patterns complete
with coloration that looks like encrusting algae. The one-
spot fringehead is named for the cirri projecting dorsally
from its head, the cirri resembling algal filaments common
in its environment (see Miller *et al.*, 1965: 57). Other
clinids and cottids have similar coloration and structures.
Such matching structures grade into *element-imitation*, considered below.

At least some flounders *(Bothus)* help further to sup-

SUPPRESSION OF OUTLINE-CONTRAST

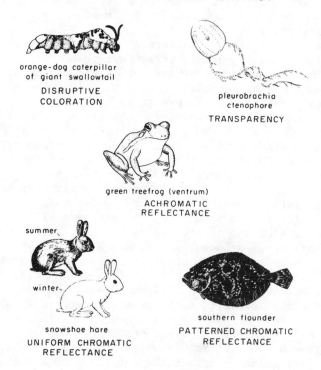

orange-dog caterpillar
of giant swallowtail
DISRUPTIVE
COLORATION

pleurobrachia
ctenophore
TRANSPARENCY

green treefrog (ventrum)
ACHROMATIC
REFLECTANCE

summer

winter

snowshoe hare
UNIFORM CHROMATIC
REFLECTANCE

southern flounder
PATTERNED CHROMATIC
REFLECTANCE

Fig 6-2. Suppression of outline-contrast by various mechanisms.

press their visual outlines by flipping sand along their
margins while lying appressed to the substrate. This
matching strategy may be called *material matching* to indicate that the animal uses materials from its environment.
Classical examples are crabs of various species that place
bits of their environments on top of their carapaces; algae,
anemones and various other plants and animals often thrive
on the crabs. Wickler (1968: 56) provides other examples.
Material matching for concealment may be difficult to distinguish from simple physical shelter, and hermit crabs
(which live in gastropod shells) are a case in point.
        Examples of these strategies for suppression of out-
line contrast--including disruptive coloration, transpar-

Table 6-II

Types of Suppression of Outline-contrast*

Type of Suppression	Adaptation(s)	Optical Effects
disruptive pattern	coloration	creates pattern irrelevant to body shape
transparency	coloration behavior	creates invisibility when: a) matches n of medium b) homogeneous irradiance c) appressed to substrate
achromatic reflectance	coloration	reflects spectrum of background in special cases
chromatic reflectance a) uniform	coloration	matches background color
b) patterned	coloration	matches background pattern
matching structure	structure	matches background pattern
material matching	behavior	adorning materials match background

---

*Adaptations grade into concealing imitations (table 6-III)

ency and various forms of matching coloration--are shown in *fig 6-2* (opposite). The principles of suppression of outline are summarized in *table 6-II* above.

Imitation

The distinction between concealment and mimicry is blurred when one considers the range of objects that animals imitate through morphology and behavior. An animal can be concealed by imitating some part of its environment (concealing imitation), or it can be conspicuous yet escape recognition by imitating some other species (animal

mimicry). Finally, an animal can act as though it is per-
forming one behavioral pattern when in fact it is perform-
ing another (feigning).

concealing imitation

Whenever an animal is evolved to resemble something
it is not, it could be said to mimic the other object.
However, "mimicry" is used both in this broad sense and
also in the restricted sense of imitating another species
of animal. To avoid confusion, I restrict the term to the
latter sense (calling the phenomenon animal-mimicry for
further clarity) and refer to other imitation as *concealing imitation*. In concealing imitation, an animal resembles some particular kind of *object* normally found in its
environment, and hence is distinguished from matching coloration (above) in which the animal merely suppresses its
tell-tale outline. Both deceptions conceal the animal because it is overlooked as an irrelevant part of the environment.

Two kinds of concealing imitation may be distinguished:
element-imitation and object-imitation. In *element-imitation* an animal resembles some specific and common object
that is one element of an environmental pattern. In patterned chromatic reflectance (above) the animal duplicates
over its surface the pattern of the environment, but in
element-imitation the entire animal becomes just one element of the visually patterned environment. The principal
examples of element-imitation are found among insects, where
shape as well as the coloration of the animal imitates some
element of its environment. Walkingsticks (Family Phasmat-
idae) and other orthopterans resemble the twigs of trees
and bushes they feed upon, the pupae or larvae of gossamer-
winged butterflies (F. Lycaenidae) resemble tree buds, tree-
hoppers (F. Membracidae) and other homopterans resemble
thorns on their host plants, and so on. Wickler (1968)
provides many examples of element-imitations (esp. pp. 50
and 60-66).

Ordinarily, behavior, physical structure and surface
coloration go hand-in-hand to produce an effective imitation. Insects are not only colored and shaped to resemble
parts of plants, but the animals also seek loci that are
appropriate to the deception. M. Itzkowitz pointed out to
me in Jamaica the habit of the elongate trumpetfish, which

swims head-downward among staghorn and other elongate cor-
als thereby becoming remarkably concealed. In this case,
it is uncertain whether the shape and coloration of the
animal are selected to be imitative, but the behavior certainly appears to be.

In *object-imitation*, the animal resembles some specific object that is not necessarily common and is not an
element of a regular environmental pattern. Object-imitation works not because the imitator blends into the background pattern, but because the viewer perceptually clas-
sifies it as something it is not. There is probably a
continuum between element- and object-imitation, but the
major distinction drawn here is that element-imitators succeed primarily when they are among the elements they resemble, whereas object-imitators succeed in deception even
when no examples of the object they imitate occur in the
same visual field of the viewer. A typical example is the
predatory leaf fishes (Nanidae) floating by in the water;
dead leaves are also imitated by many insects (*e.g.*, among
butterflies, adult *Anoea* and *Polygonia*, pupal *Limenitis*
and *Polygonia* and larval *Limenitis*; Klots, 1951). Cephalopods may hold out their arms to resemble drifting *Sar-
gassum* (Moynihan, 1975). An outlandish example of object-
imitation is the young *Papilio* and *Limenitis* larvae that
resemble bird-droppings. An avian example, which may bridge
the gap between element- and object-imitation, is the great
potoo of Latin America, which sits motionless in an unbird-
like position on dead stumps, looking much like the terminal
fragment.

There is still another kind of imitation that represents some sort of transition between resembling one element of a pattern and a unique object. Wickler (1968: 60-
64) draws attention to *imitating groups* of animals. Each
cicada of an African species in the genus *Ityraea* resembles
a flower, but the individual need not perch among flowers
to provide deception, since the insects are social and
whole groups resemble inflorescence even when no real flowers are present. Thus, each individual is an element-imitator and the entire group is an object-imitator. Wickler
(loc. cit.) also recounts the extraordinary behavior of
congregating *Tubifex*-like worms that together look like a
sea anemone.

For imitation to be effective, the deceived animal
should have no great interest in the thing imitated. Furthermore, it is necessary that the deceived animal not be

able to make a connection between the imitator and imitated that would be useful in searching for the latter. If
all stumps have potoos sitting on them, stumps become a
clue to finding potoos, so to be effective stumps without
potoos should outnumber those with potoos.

animal-mimicry

When an animal imitates another species it is usually
visually conspicuous (easily detected in the environment)
but tends to be classified incorrectly by an observer.
There is probably no firm logical distinction between object-imitation and animal-mimicry, but most cases fall naturally around one or the other of these points on a continuum. *Animal-mimicry* is therefore a special kind of object-imitation in which the "model" after which the mimic
is patterned is another species of animal. A common kind
of animal-mimicry is the "fishing lures" used by some predators, which have modified parts that resemble the prey of
a species upon which they themselves feed. The alligator
snapping turtle has an elongated dorsal projection from
the floor of its mouth that is pink and wriggles like a
worm; fish coming to investigate are quickly snapped up by
this bottom-burrowing reptile. Analogous lures are dangled
above the mouths of sit-and-wait predatory fish such as
frogfishes (Antennariidae) and batfishes (Ogcocephalidae).
There are many examples besides lures for predation
in which one part of an animal mimics another animal.
Wickler (1968: 137, figure 30) illustrates an elaboration
of the mantle of the clam *Lampsilis ovata ventricosa* that
looks like a small fish. When predatory fish snap at this
lure, the female clam squirts its larvae into the fish's
mouth, where they develop as parasites. Hence mimetic
lures are not used only for predation. An oft-cited example of mimicry is the eye-like spots adorning all kinds
of animals (see Wickler, 1968 for many examples). When
these spots occur in pairs, as on the underwings of some
moths, and when it is shown by behavioral experiments that
they frighten potential predators such as birds (Blest,
1957), they are convincing cases of *eye-mimics*. Similar
paired spots are displayed by cephalopods (Moynihan, 1975).
However, many animals are covered with spots of various
kinds, and some investigators have glibly assumed these
are eye-mimics without any real evidence. This problem

is taken up again in *ch 7*.

Entire animals may also mimic other kinds of animals, and examples are common. One species of spider I have photographed (probably *Synemosyna lunata*) is remarkably similar to an ant, and walks around with its forelegs raised like antennae, giving the impression that it is six-legged. Orchids are famous for their flowers that resemble the hymenopterans that pollinate them, so that animal-mimicry extends even across the kingdoms. Wickler (1968) relates the grisly strategy of a goby that so strikingly resembles a cleaner-wrasse it may approach a large fish with impunity and then take a bite from its flesh. Wickler's book should be perused for many other extraordinary examples of animal-mimicry.

Most examples of animal-mimicry involve similar animals, such as one insect resembling another or one fish looking like a different fish species. However, Huey and Pianka (1977) report a juvenile lizard that appears to mimic a noxious carabid beetle not only in size and coloration, but also by body posture and locomotory movement. Yet one must be ever aware of convergent evolution in postulating mimicry. It may be, as Eaton (1976) suggests, that the cheetah kit mimics the aggressive honey badger. However, the honey badger uses reversed counter-shading to be conspicuous (see next chapter) to potential predators, whereas the cheetah kit could be using the same pattern to be conspicuous to its mother or for some other reason. If the kitten lies motionless while the mother is away, its dark ventrum does not show and the young animal is concealingly colored because of matching its background. The biological literature is replete with examples of ostensible animal-mimicry, but in many cases not enough is known to be relatively sure that mimicry is really the basis of the convergence in color pattern.

Batesian mimicry is a subset of animal-mimicry, based on predator-prey interactions. This term has been applied to those mimics whose models are distasteful, poisonous, toxic or otherwise inedible or dangerous to the predator. The classical example is the edible viceroy butterfly that closely mimics the more plentiful monarch, which birds find distasteful. No discussion of mimicry is complete without mention of *warning* (or aposematic) *coloration* displayed by animals that could be harmful to a potential predator. They escape predation not by visual deception, but by "honestly" communicating the fact that they can sting or other-

wise be undesirable to other animals. Efficiency, and
hence effectiveness, of such interspecific communication
is enhanced when different species display the same kind
of warning coloration so that they are easily recognized
by predators. This evolutionary convergence to a common
type, in which every species is both model and mimic, is
called *Müllerian mimicry*. The black-and-yellow coloration
of stinging hymenopterans is a typical example, but they
in turn are mimicked by harmless dipterans, so that both
Müllerian and Batesian mimicry may occur in one evolutionary complex. A somewhat analogous example occurs among
cleaner fish that have evolved similar yellow and black
color patterns for recognition by their hosts, and in turn
the pattern is deceptively mimicked by the non-cleaner goby
mentioned above. Because these fishes are not involved
in classical predator-prey relations, the example would
not be called Batesian and Müllerian mimicry. Also, see
Wickler (1968) for the difficulties in separating these
two forms of mimicry.

A special case of animal-mimicry is *behavioral mimicry*, in which one species acts like another without necessarily being morphologically similar. The primary example
of purely behavioral mimicry appears to be the "rodent-run"
actions of some ground-nesting birds, which dart out and
run along the ground like a mouse to distract predators
away from the nest-site. Ornithologists classify this behavior functionally as a "distraction display" (Armstrong,
1964) which also covers deception that is differently based
(see next section). To return to the cleaner-mimic story,
the goby mentioned above also swims like the wrasse it resembles in coloration. The eastern indigo snake flattens
its head and neck vertically, hisses and vibrates its tail
to produce a rattling sound when cornered. This behavioral mimicry of a rattlesnake presumably is a defensive mechanism, but it is interesting to note that the indigo snake
is known to eat rattlers (Conant, 1975: 187).

Finally, it is worth pointing out animal-mimicry that
depends entirely upon emission of an optical signal. Female *Photuris* fireflies mimic the mating signals of female
fireflies of other species, attracting the males, which
they seize and devour (see Lloyd, 1977).

feigning

Imitative behavior may be part of both concealing

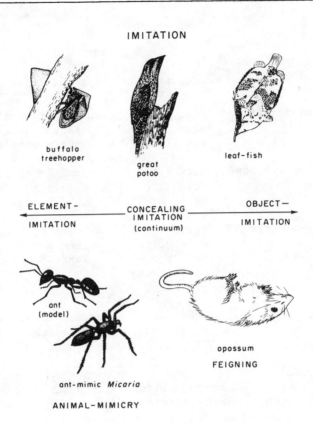

IMITATION

buffalo
treehopper

great
potoo

leaf-fish

ELEMENT-
IMITATION

CONCEALING
IMITATION
(continuum)

OBJECT—
IMITATION

ant
(model)

opossum
FEIGNING

ant-mimic *Micaria*

ANIMAL-MIMICRY

Fig 6-3. Examples of imitation, including the spectrum of concealing imitation (above) with the treehopper resembling a repetitive element in its environment (thorn) and the leaf-fish resembling an isolated object (leaf). The imitations are often more compelling in nature than shown here to emphasize the animal itself; for example, the transparent fins of the leaf-fish are virtually invisible, rendering its body shape less fish-like and more leaf-like. The ant-mimic (below) is the spider Micaria and the opossum is feigning death.

imitation and animal-mimicry, but when the animal imitates its own behavior one may refer to it as *feigning*. In feigning, the observer both detects and recognizes the animal, but may mistake its behavior.

Table 6-III

Types of Imitation

Type of Imitation	Adaptation(s)	Optical Effects
element-imitation*	structure coloration behavior	animal looks like a repetitive element in its environment
object-imitation*	structure coloration behavior	animal looks like an object that occurs in its environment
animal-mimicry	structure coloration behavior bioluminescence	animal (or part) looks like a different kind of animal
feigning	behavior	animal's behavior looks like some behavior it is not

End-points of a continuum. Also, imitating groups as a whole may exhibit object-imitation with each component animal being an element-imitator.

Feigning is a common form of deception used to distract predators from the nest (Armstrong, 1964) and hence can function like behavioral animal-mimicry. The classical case of nest-distraction is the injury-feigning or "broken-wing act" of the killdeer, in which the incubating bird runs quietly to a meter or so from the nest when a predator approaches, and then lowers one wing and drags its body in an irregular motion in a direction away from the nest. The deception distracts the predator, which may even try to catch the parent, only to have it fly off.

Injury-feigning is not the only type of behavioral deception; death-feigning is also common (e.g., Franq, 1969). The Virginia opossum is so famous for going limp when approached by a predator that its behavior has given rise to the American colloquialism *playin' 'possum*. Indeed, since movement is such a widespread characteristic of an-

imals, motionlessness *per se* is a deceptive mechanism promoting concealment, and might be classified as a special
kind of feigning. "Freezing" to remain concealed is so
common a ploy, especially among young and small animals,
that specific examples are unnecessary.

Types of imitation are illustrated in *fig 6-3* on p.
179 and the optical principles upon which they depend are
summarized in *table 6-III* (opposite).

Visual Ambiguity

Animals are detected and recognized, and their probable actions predicted, from certain visual consistencies.
Therefore, an animal can deceive by removing such consistencies or at least diminishing them, a deception I call
visual ambiguity. Naturally, almost all of the foregoing
principles of deception could be said to involve ambiguity
of some sort; the phenomena of this section are simply
those heterogeneous ambiguities that do not fall naturally
in previous categories articulated. There are at least
five kinds of such miscellaneous ambiguities: those concerning the symmetry of the animal (symmetry-deception),
instantaneous location of the animal (startle-deception),
long-term location of the animal (replicate-deception), dis-
similarity among animals of the same species (deceptive
polymorphism), and striping to conceal movement (motion-
deception).

symmetry-deception

Often an observer may predict something important
from the symmetry of the animal it sees. For example, one
expects an insect to jump away anteriorly. By creating an
ambiguity concerning the symmetry, the animal deceives the
observer. It is evident that *symmetry-deception* is not far
removed from animal-mimicry, except in this case one part
of the animal mimics another, rather than one species mimicking another. It is also the morphological analog of
feigning, in which the animal's behavior mimics another of
its behavioral patterns, once more emphasizing the spectrum
of deceptive strategies and their relationships.

Wickler (1968) provides several examples of symmetry-
deception among insects, a particularly compelling one be-

ing an unidentified lantern-fly from Thailand (p. 73, figure 14). Such ambiguity is not restricted to insects,
however, since some snakes have colored tails that appear
as heads, and this kind of deception should be looked for
among other animals. The way in which butterflies are displayed in museum collections and pictured in books--with
wings spread and flattened--draws our attention away from
symmetry-deception when the wings are folded above the
animals in their usual resting posture. Wickler (p. 76,
figure 15) pictures *Thecla togarna* at rest, with its wing
projections looking very much like head and antennae at
the rear of the animal. This butterfly also turns around
upon landing, so that the symmetry-deception is enhanced
by the behavior. It might prove interesting to look at
the landing behavior of swallowtails, hairstreaks and
other butterflies with posterior projections on their
hindwings.

startle-deception

Some symmetry-deception seems to benefit the animal
primarily by rendering its anticipated direction of movement ambiguous, and another way in which this same end is
achieved is by visually startling the observer just before
a change in direct. Such *startle-deception* seems to function by creating a visual change at the instant of directional change. However, startle-deception seems not to be
well studied, and might be a heterogeneous category of
some complexity, as two examples illustrate.

Many butterflies--*Polygonia, Cercyonis, Vanessa, Eun-
ica, Anea*--have bright and colorful dorsal surfaces on the
wings but dull undersides. Their flashy colors are exposed in flight, but then they drop into vegetation suddenly and become nearly undetectable (Klots, 1951) by holding
their wings dorsally (see *shadow-minimization* and *matching
coloration*, above). In this case, it appears as if a pursuing predator or other viewer might develop a perceptual
set (see next section) based on the exposure to the bright
coloration, and hence become confused when the pursued image no longer exists. Underwing moths and certain grass-
hoppers have similar behavior and coloration, and in grass-
hoppers a sudden sound contributes to the startle.

Ostensible startle-deception in silvery colored fishes
may depend on a different effect. The mirror-like surface

reflects light specularly *(fig 3-5)*, so that the fish is
dull at most angles but reflects a dazzling light *(ch 5)*
at the angle equal to that of the incident light. While
diving in Florida, I have noticed that when a school of
dull lookdowns turns suddenly, one sees many dazzling
flashes of light and then it becomes difficult to find
the fish immediately thereafter. This example of startle-
deception may work on combining several factors. The
flashes constitute a visual surprise for which the viewer
is unprepared, hence breaking the perceptual set and possibly frightening the viewer by means of an abrupt and not
immediately recognizable stimulus.   The stimulus may be
bright enough to dazzle the eye *(ch 5)*. Then, the brightness of the stimulus may leave an after-image *(ch 5)*, or
create a new perceptual set, either of which make it difficult to detect the now-dull fish an instant later. Finally,
the fish--having changed direction at the time of the flash--
are in an unanticipated part of the visual field, moving in
an unanticipated direction because the turn itself was
disguised visually by the flash. This example, based on
my introspective experience in the role of potential predator, accords with observations by J. Baylis *(pers. comm.)*
on other silvery fish, and also demonstrates how complex
may be the basis of startle-deception.

An apparently analogous ploy is used by the bioluminescent flashlight fish (Morin *et al.*, 1975). It swims o-
ver the reef slowly with its light on, then blinks the
light rapidly and changes direction with an increase in
swimming speed. This is just one of many uses of bioluminescence in this fascinating fish.

replicate-deception

Wickler (1968: 57-59) has drawn attention to a ploy
that allows long-term deception as to an animal's location:
the animal constructs replicates of itself in the environment. Such *replicate-deception* is known in a spider that
builds other "spiders" of its web-material and in a moth
that severs several leaves to look like its own leaf-wrapped
pupation-case. Replicate-deception works by saturating
the environment with many similar things, and the ploy
may be reversed by making similar things appear different,
as discussed in the next section.

deceptive polymorphism

All species of animals exhibit variability among individuals to some degree, and when this variability is
discontinuous the species is *polymorphic*. Polymorphic
variation in coloration is sometimes called *polychromatism*,
and many species are known to have two or more color morphs
or phases that are not correlated with sex or age. The
reasons for polymorphism are often obscure, as in the red
and gray color phases of the screech owl. In some cases,
however, an argument can be made for polymorphic coloration being an adaptation for visual deception of predators:
deceptive polymorphism. A predator that successfully takes
one prey animal may concentrate on searching for others
that look like it, and hence overlook conspecific prey that
are differently colored. Psychologists know this behavioral
phenomenon under the rubric of *perceptual set* (one tends to
see what one expects or wants to see). The ecologist L.
Tinbergen (1946) wrote of the predator having a *search image* based on previous captures.

Polychromatism of the European snail *Cepea* is one case
(B. Clarke, 1962). Presumably predators continue to hunt
for one type that therefore becomes increasingly rare until
rarity causes the predator to hunt randomly and discover a
new type. In this way natural selection maintains the polymorphism. It seems possible that the extreme intraspecific
variability of coloration found in other gastropods may be
examples of deceptive polymorphism, such as in the common
dove shell of shallow waters in Florida. Wickler (1968: 59-
60) provides several other examples in caterpillars, moth
larvae, snails and beetles.

Deceptive polymorphism is more closely related to startle-deception than one might at first think. Startle-deception depends upon rapidly changing appearance and polymorphism functions by different appearances of members of
the same species. One can imagine these strategies combined
in an animal that simply changes color from time to time in
order to prevent any observer building up a perceptual set
or seach image. Such a phenomenon might be looked for in
animals capable of facile color-change, such as fishes and
cephalopods.

Examples of visual ambiguity--including symmetry deception, startle-deception, replicate-deception and deceptive polymorphism--are shown in *fig 6-4*. There remains to
be considered, though, one last type not illustrated.

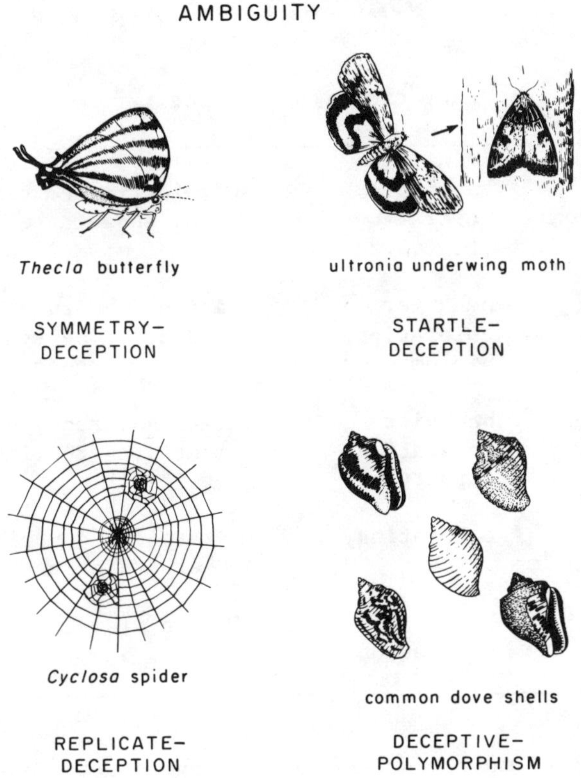

AMBIGUITY

Thecla butterfly

SYMMETRY–
DECEPTION

ultronia underwing moth

STARTLE–
DECEPTION

Cyclosa spider

REPLICATE–
DECEPTION

common dove shells

DECEPTIVE–
POLYMORPHISM

Fig 6-4. Examples of visual ambiguity.

motion-deception

Jackson *et al.* (1976) analyzed color patterns of North America snakes and found a correlation between anti-predator strategy and pattern. Specifically, species that primarily flee, rather than defensively maintain their ground, tend to be marked with longitudinal stripes. The apparent optical advantage of this pattern is that it distracts attention from cues of movement, making it difficult for the predator to track the snake. Visually, the longitudinal stripes seem stationary while the snake moves forward. This

Table 6-IV

Types of Visual Ambiguity

Type of Ambiguity	Adaptation(s)	Optical Effects
symmetry-deception	behavior coloration structure	animal oriented differently from appearance
startle-deception	behavior coloration bioluminescence	startles observer may dazzle observer disrupts perceptual set
replicate-deception	behavior coloration structure	animal saturates environment with replicates of itself
deceptive polymorphism	coloration	intraspecific variation disrupts perceptual set
motion-deception	behavior coloration	hinders visual tracking

principle of *motion-deception* also appears to be utilized by cephalopods (descriptions by Moynihan, 1975; interpretation mine). In behavior having "a strong escape component," the animals assume a posture with the arms straight out anteriorly and change into a color-pattern of bold longitudinal stripes. Another supporting case comes from open-water fishes which have no vegetation in which to hide. Due to biomechanical principles, such species tend to be long and thin (Alexander, 1967), but they also tend to be marked with longitudinal stripes, primarily for motion-deception I believe. There is a strong, albeit imperfect, correlation of habits, body-shape and pattern among fishes of coral reef areas, as thumbing a book such as Chaplin (1972) quickly confirms: those species that stay near the substrate tend to be deep-bodied with vertical bars and those that cruise over the reef in open water tend to be long with horizontal stripes. The principles of visual ambiguity are summarized in *table 6-IV*.

Combined Factors

It seems likely that most animals employ several kinds
of visual deception simultaneously, especially for advantages of concealment. In general, animal-mimicry is not
compatible with most other forms of deception because the
animals mimicked are usually conspicuous rather than concealed, and the deception succeeds not through eliminating
detection but through eliminating correct recognition.
Other combinations also seem unlikely; for example, an animal that avoids directed light is unlikely to be counter-
shaded because that coloration would be conspicuous rather
than concealing without strong vertical illumination. However, startle-deception could be used in conjunction with
animal-mimicry to frighten off a predator, as in the eye-
spots flashed by various moths. Therefore, each species
must be investigated individually to find which strategies
of deception are compatible for its particular situation.

It is worth pointing out a few effective combinations
of deception for concealment. Disruptive coloration might
be used with either achromatic reflectance or chromatic
reflectance to render one or more of its disruptive patches
visually continuous with the environment (the ploy used in
some military camouflage). Similarly, counter-shading may
utilize either reflectance-type in order to make the dorsum or ventrum blend with the environment (e.g., green dorsum of caterpillars with a white venturm). The major point
of effective combinations can be made by citing two examples where at least four principles of deception are employed simultaneously by the same animal.

Needlefishes (Belonidae) are darker above than below
(counter-shading) and the blue above tends to match the
blue of water (chromatic reflectance). Furthermore, the
silvery sides can flash light at the critical angle (startle-deception), while at other angles reflecting neutrally
the scattered ambient illumination (achromatic reflectance).

Plovers (Charadrius) also are darker above than below
(counter-shading), and their white breasts reflect ambient
light from the sand (achromatic reflectance). The piping
plover, which frequents the drier portions of sandy beaches,
and the snowy plover, which is found on sand flats and
dried alkali ponds, are both dorsally the color of dry sand
(chromatic matching). The semipalmated plover and other
species that occur on mudflats and near the water's edge
are dorsally the darker color of wet sand or mud. The

MULTIPLE DECEPTIONS

*Fig 6-5. Multiple optical deceptions for concealment ex-
emplified by plovers include counter-shading, chromatic
matching, achromatic reflectance, disruptive coloration
and substrate-appression.*

smallest of these three plovers (snowy) has a small dark
blotch, whereas the other two have a complete breast-band
(probably disruptive coloration). The larger Wilson's
plover has a noticeably enlarged breast band, and the
largest North American *Charadrius*, the killdeer, has two
breast-bands. Bold head-markings that may also be disruptive tend to follow the same trends in size. The horizontal orientation of the breast-bands may even involve
element-imitation, in that the visual field is horizontally

stratified by breaks at the shoreline and horizon. Furthermore, plovers often lie closely appressed to the beach
(substrate-appression), so that up to six different principles may be utilized by the plovers, which are illustrated in *fig 6-5*.

Finally, to re-emphasize the usefulness of studying
many species *(ch 1)*, some mention of a fascinating natural
history study of African praying mantids (Edmunds, 1976)
must be made. Some of the deceptive mechanisms found in
37 genera studied include matching coloration (different
chromatic reflectances for different backgrounds); startle-
deception; element-imitation of leaves, bark, grasses and
sticks; and animal-mimicry of ants. A particularly interesting adaptation is the molt into melanistic coloration
following a brush fire in the area, a special kind of
matching coloration that has been called "fire melanism."

Overview

Principles of visual deception may be studied conveniently in interspecific communication, where an animal
sends "*mis*information" that reduces its detectability,
recognition or predictability of behavior. Studies of
concealment and mimicry in predator-prey relations have
suggested many principles of visual deception, including
optical mechanisms for suppressing shadows and body-out-
lines to reduce detectability. There is a continuum of
deceptive strategies that ranges from coloration that
matches the background through concealing imitation of
objects in the background to conspicuous mimicry of other
species. These in turn are related to deception that creates visual ambiguities about the behavior, orientation
or location of an animal. The list of deceptive principles
provided is certainly incomplete and its organization is
tentative, but the material provides an important basis
for investigating the design of optical signals exchanged
among conspecifics.

Recommended Reading and Reference

The classic on concealing coloration is Cott's *Adaptive Coloration in Animals* (1940), available in a later
edition (Cott, 1957). Wickler's (1968) *Mimicry* makes delightful reading, and Edmunds (1974) is a recent review.

Chapter 7

NOISE

7

...overall economy...will force animals in most cases to operate their communication system as close to the ambient noise as possible. --Schleidt (1973: 375)

Most optical communication takes place in a noisy channel. This chapter concerns characteristics of signals that combat environmental noise.

Noise is entropy from any source that the receiver cannot separate from the sender's entropy, and the noise is measured by the equivocation or confusion it causes in the receiver (*fig 2-7*, p. 50). It is useful to distinguish two kinds of optical noise: that which physically changes the signal during transmission (which may be called *transmission noise*), and that which overwhelms the signal with extraneous entropy (*detection noise*). Shannon's treatment of noise (Shannon and Weaver, 1949) is directed primarily to electrical interference in telecommunicational channels with incessant transmission, where the distinction between transmission and detection noise is generally unnecessary. However, one can imagine interference changing a telegraphy dot sent to a dash received (transmission noise) or adding dots and dashes when no one was pressing the sending key (detection noise). In optical communication, transmission noise is caused by materials such as fog, turbidity or plants between sender and receiver, and detection noise occurs when the receiver cannot distinguish a signal from its optical background.

My approach to optical noise first considers principles of conspicuousness, many of which may be deduced

-191-

by "reversing" the principles of deception articulated in
the previous chapter. Then I search for how these principles may be manifest in different environmental circumstances by considering differences in available light,
differences in optical background and differences in the
clarity of the medium--in each case attempting to identify how the environment structures optical signals to increase the signal-to-noise ratio. Finally, I consider
the problem of how an animal might remain concealed from
predators and other observers while being conspicuous to
its social companions with which it is communicating.

At this point the emphasis changes from summaries
and reorganizations of facts from the literature to ex-
plorations and hypotheses of the design of optical signals. In this sense, the chapter extends the kind of
treatment given "pseudosignals" *(ch 4)* and deception *(ch
6)* to optical social signals, trying to identify factors
that structure animal behavior and morphology. Where possible, I use an informal comparative method *(ch 1)* to
check predictions from the hypotheses developed, but these
checks should be taken for just that: preliminary assess-
ments to see if the hypotheses are worth mentioning.

It is useful to keep in mind the diversity of elements that may be part of an optical signal: orientation,
movement and shape as behavioral elements *(fig 4-1)* and
reflecting surfaces and structures *(table 4-III*, p. 105)
as morphological elements. Signal-to-noise considerations
seem primarily to involve adaptations of reflecting surfaces, where the most important variables are hue, saturation and brightness *(table 5-I*, p. 138), along with size,
shape and type of reflectance *(table 4-III)*. Furthermore,
animal coloration cannot be predicted purely on the basis
of signal-to-noise considerations, for much of animal coloration is dictated by needs of concealment *(ch 6)* or other
factors *(ch 4)*. Even if these were accounted for, the
framework remains incomplete without scrutiny of how the
information imparted by the signal helps to structure its
characteristics, the problem taken up in *ch 8* below. Therefore, in making the informal checks of hypotheses developed
in the present chapter, I have tried to find species in
which other factors are minimized so that the direct effects of the signaling environment on the characteristics
of signals become the chief factor. At this stage of
knowledge such oversimplification seemed better than no
check at all: it is a crooked wheel, but the only wheel
in town.

Principles of Conspicuousness

The overall strategy for combatting optical noise is to utilize signals that can be discriminated from other stimuli in the environment and that are environmentally robust in the sense of being discriminable even when altered during transmission. This combination of desirable characteristics may be called *conspicuousness*, and the point of this chapter is to discover what constitutes conspicuousness. To do this, I try first to identify the general principles of conspicuousness (this section), and then to search for their manifestations under different conditions. For example, color-contrast is a general principle of conspicuousness, but it is manifest differently in different environmental circumstances: a male cardinal may be conspicuous against green leaves but not among red flowers.

Much of the visual deception practiced by animals serves the function of concealment *(ch 6)*, so by reversing principles of deception one should be able to deduce some principles of conspicuousness. The previous chapter identified suppression of motion, shadow, shape, contrast and other regularities as elements of deception, so this section investigates enhancement of these optical factors as elements of conspicuousness.

movement

Movement is the most obvious principle of conspicuousness. When an object moves within a stationary visual field, it sets itself off immediately. There are even special cells in the visual system of some animals that respond only to certain kinds of movement within their receptive fields *(ch 5)*. Movement is a chief component of intrinsic optical signals *(fig 4-1)*, and along with body shape (including expressions) has received the most attention by ethologists (Hailman, 1977a). The task is therefore not to document the importance of movement in conspicuousness, but rather to see if certain attributes of movement are especially conspicuous.

The first point is that movement must be within a certain range of *speed* in order to have maximum conspicuousness. Very slow movements (*e.g.*, locomotion of sloths) will not be detected readily, and very rapid movements

may exceed the critical fusion frequency of the eye (fig
5-13). There is a broad range within these limits in
which movement will be easily detected, and since certain
visual cells appear to have optimum detection at certain
speeds, it seems possible that there are optimum speeds
of movement for conspicuous optical signals. For example,
parent gulls move their bills in front of their young
chicks, and if hungry, the chicks beg for food by pecking
at the bill. Chicks have a preferred speed of movement,
found in experiments to be about 12 cm/s, and the mean
speed of the parental bill in the signal-movement measured
from movie film was 14.5 cm/s (Hailman, 1967a). The optimum speeds may be relative to the size of the object.
For example, Schleidt (1961) found that turkeys discriminate flying predators from other entities overhead by
the number of body-lengths the hawks move per unit time.
Discrimination is thus independent of distance from the
hawk and is coded more complexly than simple speed of the
image across the retina.

Directionality of movement is a second variable that
deserves experimental attention. For example, the parent
gull's bill is held vertically and moved horizontally in
the usual signaling situation. When a horizontally held,
vertically moved bill is presented to chicks, their response rate is much lower (Hailman, 1967a). In fact, experiments using models of both orientations, as well as
motionless and moved in both directions, yielded significant interactions among the variables. The results suggested that movement across the long axis of the bill and
movement parallel to the visual horizon both play a part
in the characteristics of the optimum signal.

An obvious principle of conspicuousness in movement
is repetition. Repeated movements are a form of temporal
redundancy (ch 2) and are exhibited by many animals. For
example, the male turkey performs strutting movements
throughout the day (Schleidt, 1964). The tail-flashing
of the dark-eyed junco is more eye-catching as a repeated
spreading of the tail to flash the white outer feathers
than would be a chronic spread. Besides adding conspicuousness, of course, repetition of a signal and the in-
terval between signals can be used to encode specific
information (ch 8).

Sudden onset of a signal also makes it conspicuous
by virtue of emphasizing the change from motionlessness
to movement. This principle is used protectively in star-

---

tle-deception *(ch 6)*, where conspicuousness is followed
by immediate concealment as a mechanism promoting confusion. In social communication, sudden onset followed
by other elements of conspicuousness make a signal more
obvious. Simpson (1973) emphasizes the role of surprise
in conspicuousness, a notion similar to that of "news"
mentioned in *ch 2*.

A full classification of movements and the variables
that make them especially conspicuous has not been devised,
but it seems likely that other factors contribute. In repeated movement, *rhythmicity* may promote conspicuousness
to visual systems particularly tuned to certain frequencies, whereas in other cases arhythmic, jerky movements
may be more conspicuous, as in claw-waving displays of
fiddler crabs (see figure 21 in Hailman, 1977a). Indeed,
if visual systems are tuned to particular speeds, directions and frequencies of movement, generally low variance
in repeated performance—movement *stereotypy*—is a factor
of conspicuousness itself (Marler, 1974: 36). More is
said about stereotypy in *ch 8*, below.

These sorts of variables in movement must also be
considered relative to the environmental context in which
the motion occurs. A leaf fish *(fig 6-3)* moves, but is
concealed because it moves by drifting with the current.
Therefore, conspicuous movement must be at a different
speed or in a different direction from movements of the
relevant optical environment. Movement in relation to specific environments is considered later in this chapter.

Finally, movement may be exaggerated by extrinsic
optical signals. For example, the branch-shaking of many
monkeys creates a far more conspicuous threat display than
would movement of the animal alone. Many kinds of movements carry specific information (next chapter) and hence
are structured for other than mere conspicuousness, but
most movement helps   render the sender more conspicuous.
Indeed, Noble and Curtis (1939) felt that the primary role
of signal-coloration in the jewel fish was to accentuate
movement, so I now turn to conspicuousness of coloration.

reversed counter-shading

As noted in *ch 6*, animals cast a shadow upon the substrate and also shadow the lower parts of their own bodies.
I know of no social signals that utilize the substrate

shadow. An animal could, however, become visually conspicuous by emphasizing its body-shadow through *reversed
counter-shading*: the "with-shadow" pattern of Albrecht
(1962) or "inverse counter-shading" of G.W. Barlow (1974).
The previous chapter mentioned examples of animals that
possess reversed counter-shading because they habitually
pose upside-down, thus making the pattern concealing.
There are, in addition, animals leading rightside-up lives
that are light dorsally and dark ventrally, and hence
quite visually conspicuous. A striking example is the
male bobolink, an icterid blackbird of hayfields that is
largely white and cream-colored above. Few animals are
chronically counter-shaded in reverse, although in male
eider ducks *(Somateria)* and among fishes in some of the
grunts (Pomadasyidae) the pattern is permanent. More u-
sually, reversed counter-shaded plumages like that of the
male bobolink or certain plovers *(Pluvialis)* are assumed
only for the duration of the breeding season. In fishes,
the coloration may be assumed for a shorter period of
some specific behavior requiring conspicuousness, as in
parental defense (Keenleyside, 1972; Baylis, 1974a); see
also Morris (1958).

More common than reversed counter-shading of the entire animal is reversed counter-shading on a local part
of the body used in display. Among birds, for example,
several shorebirds assume black throat, breast or belly
patches for the breeding season (*e.g.*, ruddy turnstone,
rock sandpiper, dunlin) and in several orioles *(Icterus)*
the black throat and upper breast permanently contrast
with the orange crown and neck. Indeed, there are so
many cases of breeding plumages of birds and courtship
colorations of fishes that involve at least small areas
of dark coloration below lighter coloration that one suspects reversed counter-shading is an extremely widespread
attribute of optical signals in animals. Some examples
of reversed counter-shading are shown in *fig 7-1*.

Hamilton and Peterman (1971) discuss counter-shading
in coral reef fishes, and if I understand their point
correctly, emphasize the temporal contrast in coloration
when a fish changes rapidly from the concealingly counter-
shaded pattern to display coloration. The display coloration itself, in the examples discussed, is not counter-
shaded in reverse, but derives its conspicuousness by contrast with the counter-shaded pattern, so that the latter
is selected for because of its role in signaling. They

REVERSED COUNTER-SHADING

♂ spectacled eider

Spanish grunt

summer   winter

♂ hooded oriole

♂ bobolink

Fig 7-1. Some examples of reversed counter-shading that render animals visually conspicuous. Some animals are wholly reverse counter-shaded or nearly so, whereas others, such as the male hooded oriole, have restricted regions of conspicuous shading. Male spectacled eiders are habitually shaded whereas male bobolinks are shaded only during the breeding season, and some fishes may intensify shading for relatively brief periods.

state that "countershading hides a signal rather than the fish" (p. 363), although it is not clear to me how it does one without doing the other.

enhancement of shape

The third general principle of conspicuousness is *enhancement of shape*. In this case, however, conspicuousness of signals cannot be deduced by simple reversal of concealment of entire animals. Animal outlines are suppressed by mechanisms such as transparent appendages or disruptive coloration *(ch 6)*, but opaque appendages and non-disruptive coloration can hardly be considered specific adaptations for conspicuousness. One kind of non-disruptive coloration--*uniform coloration*--is indeed conspicuous in providing a clear appreciation of an animal's outline, when the color contrasts with the background (see below on *surface-contrast*). Thus, the all-blue male indigo bunting, all-red cardinal, all-black crow and other uniformly colored animals provide good visual outlines. Another method of enhancing shape is to *outline* or edge it in a contrasting border, as in the monarch and other butterflies. This strategy is not used often in vertebrates, but certain fish having distinctive fin shapes may emphasize the shapes with black edging (*e.g.*, the palometa). T. Johnston (*pers. comm.*) has suggested that depth perception allows a solid, three-dimensional animal or object to be detected more readily than a flattened, two-dimensional one. Therefore, one might expect flattened body parts (lepidopteran wings, fish fins, *etc.*) to show bordering contrast more than other body parts. Furthermore, a flat surface has an unambiguous border that can be outlined, whereas a solid object may be seen from different angles and hence would favor enhancement by uniform coloration.

Enhancement of outline is more commonly employed locally on an animal's body to emphasize the shape of a signal-patch. Here, black edging of a shape is commonly employed, as in the edging of the golden-crowned sparrow's and the golden-crowned kinglet's golden crowns, the blue jay's and barn owl's white faces, the hooded merganser's white crest, and so on in many birds. Fishes commonly employ the same principle of outlining a signal-patch in black, as with the gold spots on the goldspotted eel, but more often have a dark spot outlined in white or other light color, as in certain frogfishes *(Antennarius)*, young of certain damselfishes *(Eupomacentrus)* and butterflyfishes *(Chaetodon)*, young rock beauty, silver porgy and many others. Birds sometimes have signal coloration outlined in white (*e.g.*, the russet belly-patch of the gray partridge), but convincing examples are fewer. Examples of both polarities are common among lepidopterans. As in the case with local reversed counter-shading, local outlining of shape in less dramatic examples is probably widespread.

Another mechanism for enhancement of outline is to create a signal-patch of *unusual shape* so that it is conspicuous by virtue of its novelty. This suggestion is simply a manifestation of the general principle of surprisal or news *(ch 2)*. The strategy of unusual shape, however, is difficult to identify in practice. Animal body shapes may be changed by disguising appendages or by other deceptive means that obliterate tell-tale shapes *(ch 6)*, but it is difficult to reverse this principle because unusualness of shape is hard to specify in the abstract. One attribute of unusualness of shape in nature might be *geometric regularity*: for example, circles, triangles and rectangles are rare both as body shape or as any other pattern in nature—except as optical signals. Eyes, of course, are circular, and round patches of animal coloration are often explained as eye-mimics *(ch 6)*. Nevertheless, round signal-patches seem far too widespread to be explained simply as eye-mimics, and one element promoting circular patches may be the unusualness of their shape. Many examples from fishes were cited above.

Triangular patterns are surprisingly recurrent in avian plumages, particularly as color-patches assumed by the male for the breeding season. Some display triangles are nearly equilateral, as in the white triangle of the hooded merganser's crest (when fanned), the paired triangles (dark or light) in tails of several ptarmigans *(Lagopus)*, the black triangular patches on the wing-tips of the black-legged kittiwake, or the russet facial patch of the Cape May warbler. Other triangles may be long and thin, but still strikingly noticeable in shape, as in the black wing-marks of Sabine's gull, white wing-marks of Bonaparte's gull, mustache marks (red or black) of the male common flicker, the red crest of the pileated woodpecker, or the black facial marks of golden-winged and olive warblers.

Other regular geometric shapes, like rectangles, as well as certain improbable shapes, may also be selected for as signals because of their unusualness and hence compelling visual conspicuousness. Rectangles, or parallelograms, with quite straight edges occur in the speculum patches on the wings of many ducks (see figure 20 in Hailman, 1977a), but are not otherwise common in birds. Similarly, a marine fish, the black margate, possesses a compelling black parallelogram on its ventro-lateral sur-

face. Quite improbable shapes of coloration are found on
some birds--*e.g.*, the white crescent on the male blue-winged
teal and male Barrow's goldeneye (see figure 9 in Hailman,
1977a)--but as noted above, the notion of unusualness is a
difficult one outside of regular geometric shapes known to
be rare in nature. Some examples of enhancement of shape
are shown in *fig 7-2*.

ENHANCEMENT OF SHAPE

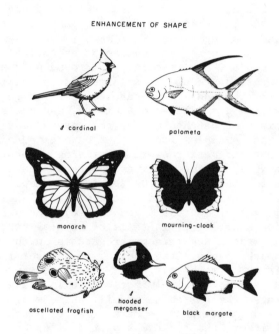

♂ cardinal   palometa

monarch   mourning-cloak

oscellated frogfish   ♂ hooded merganser   black margate

*Fig 7-2. Enhancement of shape exemplified by various an-
imals. An animal may be uniformly colored, like the cardinal, to contrast with its background, or may outline critical (particularly flattened) parts with special coloration, like the palometa. Light colors may be outlined in
black (monarch) or dark outlined in white (mourning cloak).
Conspicuousness may also be achieved by shapes that are
infrequent in nature, such as geometrically regular shapes:
circles (frogfish), triangles (merganser) and rectangles
(margate).*

Extrinsic signals may also utilize unusual shapes to increase conspicuousness. For example, the bower of *Amblyornis macgregoriae* is a triangular pile of sticks surrounded by a circular ring of moss on the ground (Firth, 1970). Other species of bowerbirds have even more elaborate bowers, but many have triangular and circular elements.

surface-contrast

An obvious principle of conspicuousness is *visual contrast*, but the question is: what is contrast, how is it measured and what does it predict about signal-coloration? *Chapter 5* divided apparent surface coloration into the phenomenal dimensions of brightness, hue and saturation, and developed the colorimetric notions of luminance, dominant wavelength and excitation purity as methods for measuring the phenomenal dimensions *(table 5-I)*. Consider each in turn as a component of visual contrast.

The most straightforward kind of contrast depends simply upon the amount of visible light reflected from two surfaces: *brightness contrast*. The extreme brightness contrast occurs when one surface reflects nearly all incident light and the other reflects almost none of it, so that the surfaces appear achromatically white and black, respectively. Brightness contrast is employed commonly in emphasis of shape by outlining an animal or signal-patch in either black or white (see examples cited in the previous section). Furthermore, brightness contrast is also employed in homogeneous coloration to emphasize animal shape, which may partly explain the black or dark gray coloration of various birds that live in open, well-lighted habitats such as fields and marshes (*e.g.*, crow, blackbirds, juncos, male lark bunting, male phainopepla, *etc.*).

An animal may use brightness contrast for other than outlining a signal-patch or emphasizing body shape through homogeneous coloration: signal-patches themselves may be black or white. One may expect achromatic coloration of signal-patches anywhere the background coloration of the patch is rather light or dark. For example, the black male lark bunting possesses boldly contrasting white wing-patches and the nearly all-black white-headed woodpecker has, of course, a contrasting head. Or, reversing the polarity of contrast, the snowy owl is marked with black scalloping and black about the face, terns (Sterninae)

of many species are white or light gray with contrasting
black caps, and the wood stork is white with a dark gray
head and black wing-patches. Most signal-spots on otherwise silvery fishes are contrastingly black. Similarly,
black patches and spots that are presumed social signals
may be found on birds having chromatic but quite light
ground colors such as the male dickcissel's black throat
on its yellow breast, the male cardinal's black face on
its red body, and so on. Of course, black markings on
birds likely have other functions as well: resistance to
abrasion, reduction of glare, *etc.*(see *ch 4*).

The second dimension of visual contrast is that of
hue, correlated primarily with the spectral peak of reflected light *(table 5-I)*. Before attempting to define
contrast in hue *per se*, however, it is worth noting that
differences in hue may actually operate as brightness contrast in special cases. The most convincing example comes
from plant-animal communication *(fig 2-2)*, where it has
been observed since the turn of the century that humming-
bird-pollinated flowers are overwhelmingly red. This is
not due, I think, to any ancestral predisposition for
hummingbirds to be attracted to red, although a red-pref-
erence could have evolved subsequently. Rather, flowers
adapted for hummingbird pollination must produce large
quantities of nectar to be attractive to a large animal;
therefore, it is not adaptive for such flower to attract
insects that might deplete the nectar or pollen without
effecting pollination. Because insects, by and large,
have spectral sensitivities shifted toward shorter wavelengths *(ch 5)*, they see well in the UV but poorly in the
"red" part of the spectrum, where vertebrate sensitivity
is generally good. Therefore, a red flower looks bright
(and hence contrasting with its background) to a hummingbird, but dark (and hence not so contrasting) to an insect:
red is the only color that is conspicuous to birds without
being so to insects (Pijl and Dodson, 1966; Raven, 1972).

Contrast of hue in a more general sense is difficult
to define. Non-uniformity of spectral discriminability
defines phenomenally similar blocks of wavelengths in the
visible spectrum *(fig 5-10,* p. 133), but does not reveal
whether blue is equally discriminable from, say, red and
yellow. It seems intuitively likely that complementary
colors provide the best contrast in hue. Complementary
wavelengths are those monochromatic lights that combine
to yield an achromatic sensation. *Figure 7-3* plots rep-

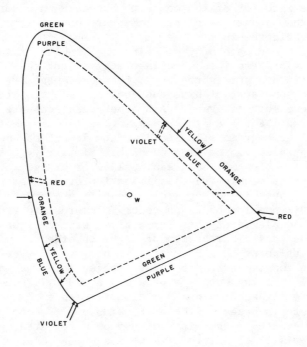

*Fig 7-3. Complementary dominant wavelengths may be de-
fined by the C.I.E. chromaticity diagram for equal-energy
white illumination. (See fig 5-12 on p. 137 for explanation of the diagram and the scale of wavelengths.) Hues
have been labeled on the outside periphery of the spectral
locus, with approximate hue-boundaries marked by arrows
pointing toward the white-point (W). When extended through
the white-point, the arrows define boundaries of complementary hues on the inside periphery of the diagram (broken arrows). Therefore, complementary hues are found on
either side of the spectral locus (unbroken line), so the
diagram may be read from inside-out or outside-in. For
example, beginning at the lower left, the complement of
violet is always a green, but the complement of blue may
be a green, yellow or orange. Exactly the same relations
may be found at the opposite side of the diagram.*

resentative boundaries of hues on the outside of the spectral locus of the C.I.E. chromaticity diagram. A line has
been extended from each boundary through the white-point

to the interior of the opposite boundary in order to define the boundaries of complementary hues on the interior
of the diagram in *fig 7-3*.

The approach of *fig 7-3* to the problem of complementary colors provides some instructive cautions. Some com-
plements are straightforward: the complement of yellow is
always some sort of blue and that of red, violet and purple some sort of green. All other relations are more complicated. For example, depending on the dominant wavelength of a green stimulus, its complement may be orange,
red, purple, violet or even blue. Similarly, the complement of a blue may be green, yellow or orange. *Figure 7-3*
shows that meaningful simplification of the notion of complementary hues is not possible, despite statement to the con-
trary in popular books on color. In order to find the "i-
deal" complementary hue for a given stimulus, one must know
the spectral distribution of light in order to plot the
stimulus-point on a chromaticity diagram for calculation
of the complementary dominant wavelength. One should also
recall that the entire approach via the C.I.E. chromaticity
diagram is based on human vision, and that the particular
use here is based on the white standard of equal-energy.
The approach is therefore of use only so far as animal
color-vision of interest resembles human color-vision in
metameric properties (see *ch 5*), and when the relevant
white-point can be determined as applicable to a given perceptual situation.

The final component of visual contrast is *saturation,*
the dimension of surface coloration given least attention
in the perception literature. Like brightness, the contrast between two surfaces differing in saturation is
greatest when the saturations differ by the maximum amount,
the range being from achromatic to monochromatic *(ch 5)*.
Only one useful generalization seems extractable: since
there are few highly saturated stimuli in the natural environment other than flowers (which are actually optical
signals of plants to animals), one may expect animal signal coloration to be highly saturated. Because of the
broad bands of wavelengths reflected by any non-fluorescing
surface that is not dark, reflecting surfaces can never be
really monochromatic. Nevertheless, within the constraints
provided by biochromes and schemochromes *(ch 4)*, one expects "pure" colors rather than pastel shades from animal
signals. This expectation so obviously obtains in birds,
fishes and many other animals that an explicit check for
reasonableness is unnecessary.

The problem, then, is to put contrast of brightness, hue and saturation together in a framework that integrates these simultaneously occurring properties of surface coloration. Unfortunately, no ideal color-space has yet been devised, as noted in *ch 5*. The model required is one of a three-dimensional space in which one distance between two plotted stimuli expresses the contrast between them. Ideally, the mode should be based on empirically derived functions of discriminability over the three-dimensional space, but I have been unable to find such data, even for human perception. Various color-spaces have been proposed, as mentioned in *ch 5*, but each has certain drawbacks, so that a coherent and predictive model of surface-contrast remains to be devised; one attempt is mentioned at the end of this chapter.

Finally, one may note that the conspicuousness of bright, complementary, saturated colors is not restricted to intrinsic signals. Bowerbird decorate their bowers with colorful berries, flower petals, leaves and other objects they can find (Marshall, 1953; Gilliard, 1969).

miscellaneous principles

A few factors promoting conspicuousness do not fall readily into the above categories: image-size, repetition of coloration in space and signal-rarity.

It is obvious that, in general, the greater the *image-size*, the more conspicuous an object--at least up to a point. Size of an animal is due to factors other than image-size in optical communication, to be sure, but animals can vary the imaginal areas of their optical signals by the distance at which they display from intended receivers and by the size of any signal-patterns of coloration. On expects relatively large color-patterns to be correlated with long-distance signaling, as in territorial advertisement. Short-distance communication, as in precopulatory display, might involve smaller movements or color-patches. The prediction is difficult to check without detailed investigation of the communication in the species of interest.

Size of an image has upper limits, however. If signal-conspicuousness depends in part on visual contrast with the background, then one predicts that there should be some upper limit to the size of a signal. A case can be

made for this principle in the contrasting red color-patch
on the bills of parent gulls, which are signals that e-
licit and direct the begging responses of their chicks
(e.g., Hailman, 1967a). Species with relatively thin bills
have entirely red lower parts, whereas in species with
thick bills the red signal-patch is confined to the bill-
tip (where mandibles converge to a narrower width) or to
a red spot on the lower mandible, as shown in *fig 7-4*.

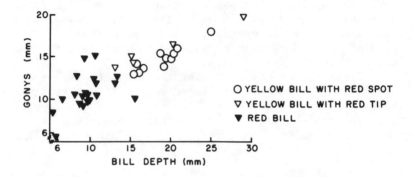

*Fig 7-4. Optimum size of signals exemplified by bills of
gull species. Small bills are wholly red, but larger species have yellow bills with red restricted to the narrower
tip or to a narrower spot on the lower mandible (after
Hailman, 1967a).*

Repetition in space is similar in effect to temporal
repetition. Both principles of conspicuousness derive
their advantage from redundancy *(ch 2)*. Spatial repetition works by creating a regularity that is uncommon in
nature and hence attractive to the eye; in this sense, it
is related to unusualness of shape (see above). Repetitive patterns are often found in birds, as a series of
tail-spots in cuckoos *(Coccyzus)* and trogons *(Trogon)*, a
series of head-stripes in sparrows *(e.g., Zonotrichia)* and
certain wood warblers (Parulidae), repetitive wing-bars,
breast-spots, breast-stripes, *etc.* The principle is no
less common among fishes, with repetitive spotting in such
species as the great barracuda, graysby, spotted goatfish
and others; repetitive vertical barring in schoolmasters,

SPATIAL REPETITION

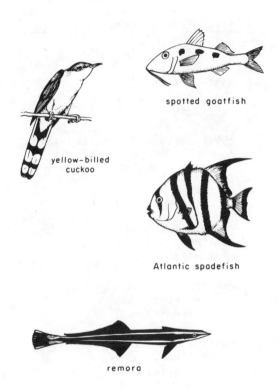

Fig 7-5. *Spatial repetition of coloration, as in white spots on the cuckoo's black tail, black spots on the goatfish's silvery body, black bards on the spadefish's light gray body or white stripes on the remora's dark body.*

spadefish, banded butterflyfish, sergeant major and others; repetitive horizontal striping in sharksucker, several grunts *(Haemulon)*, slippery dick, several parrotfishes *(Scarus)* and others. Some examples of spatial-repetition are shown in *fig 7-5*.

Finally, the notion of surprise in signals, mentioned above (see also Simpson, 1973) was developed earlier by Moynihan (1970) into a general evolutionary theory of displays. Noting that rare signals individually transfer more

Table 7-I

Some Principles of Visual Conspicuousness

Category	Specific Manifestations
movement	optimum speed (absolute and relative to image-size) optimum directionality (relative to shape and environment) repetition sudden onset rhythmicity (regular and jerky)
reversed counter-shading	
enhancement of shape	uniform coloration outlining border unusual shape
surface contrast	brightness contrast complementary hue high saturation
miscellaneous	optimum image-size spatial repetition signal-rarity

information than common ones (the notion of "news" discussed in *ch 2*), Moynihan points out that *signal-rarity*
is a principle of conspicuousness. He goes on to speculate that being conspicuous and effective, such rare signals will be selected for during phylogeny, but in becoming more common lose their rarity and hence effectiveness,
and so eventually disappear from the repertoire. That
phylogenetic theory is not the concern here, but it may be
noted that it is a difficult theory to evaluate empirically.

The principles of conspicuousness that I have been
able to identify are summarized in *table 7-I*. No doubt
the list is incomplete, but it does serve as a starting
point for attempting to see how specific conditions of
the communicating environment may help structure the
characteristics of animal signals.

Available Light

The signal as received depends upon four primary factors: the light falling on the sender, the behavior and morphology of the sender, the background against which the sender is observed, and the medium between sender and receiver. The strategy of analysis is to see how the behavior and morphology of the sender are affected by the other three variables so as to make the sender's signals as conspicuous as conditions allow. This section begins by investigating differences in light available under different environmental circumstances. The problem of available light may be decomposed into four general situations: nocturnal light, sun-angle (including time of day, time of year and latitude), altitude above sea-level and general habitat.

extremely dim illumination

In most cases the nocturnal environment is not devoid of ambient light in the way in which a deep cave is, for instance. Experienced campers know that starlight alone on a clear night provides the human eye with considerable information, and full moonlight seems bright to the dark-adapted eye. Because such nocturnal ambient light is below photopic operating levels of our cones *(ch 5)*, we see the world in black and white. Based on this introspection we expect nocturally signaling animals to have primarily white signals; any other color would reflect less light and hence reduce the brightness contrast. It is not surprising, then, that nocturnal birds such as most goatsuckers (Caprimulgidae) have white patches on the tail, neck or wings, surrounded by black to enhance the contrast. In interkingdom communication, moth- and bat-pollinated flowers are white. It seems worth pointing out, however, that there has been no systematic attempt to show empirically that animals in general resemble us in lacking color vision at nocturnal levels of ambient light. Data such as the presumed rod/cone break in dark-adaptation curves *(fig 5-4,* p. 124) do suggest that nocturnal color-vision is unlikely, but the evidence is scarce and there seems to be no theoretical argument compelling such belief.

The alternative to white, maximumly reflecting coloration is of course bioluminescence. As expected, deep-sea and nocturnal organisms have been the primary animals to evolve bioluminescent social signals.

sun-angle

Everyone knows from experience that the spectral distribution of direct sunlight differs through the day: sunlight is quite white near midday and orange as sunset approaches. (Early risers know the morning sunlight is similarly orange, but I have only rare personal confirmation.)
The change in light quality is due to Rayleigh scattering
(eq 3.9), which is greatest at lowest angles of the sun,
when sunlight must travel a longer path through the atmosphere *(ch 3)*. For a given time of day and year, the sun's
angle also decreases as one moves along a line of longitude
from the equator toward a pole on the earth's surface;
latitudinal differences in ambient light are due to the
same principal factor as hourly differences. Finally, because the earth is tilted on its axis, the north pole is
angled farther from the sun than is the south pole in De-
cember, whereas in June the relations are reversed because
the earth has traveled half its orbit around the sun while
maintaining the same tilt on its axis. Seasonal differences in ambient light quality are therefore also due to
sun-angle. The interaction of time of day, latitude and
season in determining sun-angle may be shown graphically
as in *fig 7-6*, which simplifies the problem of estimating
general quality of ambient light.
Two important aspects of available light change with
sun-angle: as the sun becomes higher in the sky, the absolute irradiance falling on the earth increases and the
spectrum shifts. *Figure 7-7* on p. 212 shows the major
effects, simulated by the SOLREF program of Dr. W.P. Porter for Death Valley, California on 21 June (summer sol-
stice). Just after sunrise (0530 hours, local standard
time), when the sun's angle is less than 5° above the horizon, the spectrum is a monotonically increasing function
of wavelength. A half-hour later the total irradiance is
much higher and the spectrum has flattened to a considerable extent. An hour after this there is still some increase in irradiance-level, but the spectrum has changed
very little. After five more hours a similar increase in
irradiance has occurred, but again, little change in the
spectral distribution of light. The changes following noon
are symmetrical with the morning functions. From this
simulation, one can see that the primary effects of color-

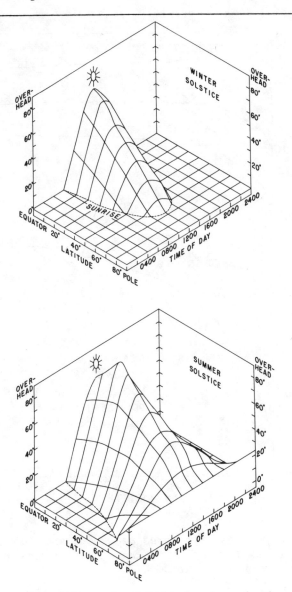

Fig 7-6. The sun's angle (altitude above horizon) depends upon the time of day and the latitude, as shown by the three-dimensional plots. Furthermore, these relations change through the seasons, as shown by differences in plots for the winter and summer solstices.

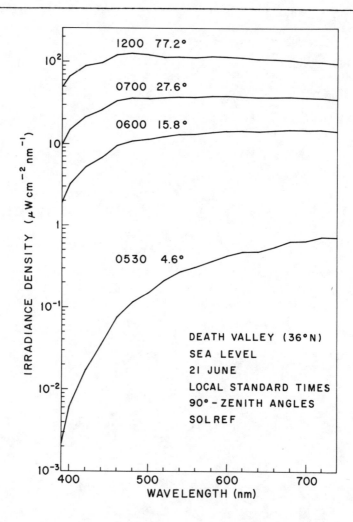

*Fig 7-7. Spectral irradiance depending upon sun-angle,
exemplified by time-of-day changes for one date at one
location (computer simulation).*

quality (spectral distribution) occur only very near sunrise and sunset; *i.e.*, at quite low sun-angles. Because
of details not important here, the latitudinal effects are
slightly different, but in general only the lowest sun-
angles (say, less than 10°) have important effects on color-

quality of ambient irradiance, making it decidedly orange
in appearance.

The angle from which an object such as an animal is
viewed relative to the sun's rays makes a larger difference in the object's appearance when the sun-angle is low
than when it is high. At low sun-angles a viewer with his
back to the sun sees an animal illuminated by the orange
sunlight, and if the animal is against the sky, that sky
will be intensely blue because of Rayleigh back-scattering. Looking toward the sun, however, the animal will be
illuminated primarily by dimmer back-scattered plus reflected light, the former being bluish in emphasis, the
latter being reddish or otherwise colored depending upon
the reflecting surfaces in the neighborhood. At high sun-
angles, on the other hand, the animal will be similarly
illuminated by direct and indirect sunlight whatever the
angle of observation.

It seems likely that these differences in ambient
light that result from differences in sun-angle will have
some influence on the evolution of signal-coloration. In
particular, an animal that signals at low sun-angle may be
expected to utilize long-wavelength-reflecting signals (red,
orange, yellow) when signals are colored rather than being
achromatically white. This expectation is based on the
fact that total irradiance is relatively low, so that a
bright signal is conspicuous. Signals that reflect most
strongly in the part of the spectrum where ambient light
levels are high (*i.e.*, long wavelengths) will appear brightest. Furthermore, against blue sky or green foliage, such
long-wavelength-reflecting signals will show high color-
contrast (*fig 7-3*, p. 203).

The difficulty of checking these expectations concerning signal-coloration used at low sun-angles resides in
lack of information about the timing of animal communication. Most temperate and polar animals breed during the
summer, when the sun-angles at their latitudes are relatively greatest.

A body of information available, however, is latitudinal distribution of birds, a major group employing
optical signaling. One expects that the more northerly
the range of a north temperate species, the more likely
it is to show red or other long-wavelength signal coloration. A comparative check may be made by holding phylogenetic relationships relatively constant and looking for
colorational differences that correlate with latitude a-

mong species within the same taxonomic group.

Species of the Subfamily Carduelinae range over a
large latidudinal spectrum. In North America, the most
northerly finches of this group are the redpolls (Acan-
this), which breed in Canada and Alaska and range south
in winter only to the northern tier of American states.
The primary signal-coloration of these northerly birds is
a bright red cap; there is also black around the bill, and
the hoary redpoll has a white rump. Nearly as northern
in distribution are the crossbills (Loxia) and the pine
grosbeak, in which the males are rosy to orange-red dorsally. The purple finch and its relatives (Carpodacus)
are more southerly and have the same male red coloration,
but these birds also frequent the mountainous areas of
western North America where the added problem of altitude
occurs (see next section). The most southern carduelines
are the goldfinches (Spinus), which all range into Mexico;
only the pine siskin from this group breeds a significant
distance north into Canada. All these birds lack red entirely, and have signal-coloration mainly of yellow and
black. Roughly the same relations apply in Europe, except that the European goldfinch ranges very far north
into Scandinavia, and unlike the southerly New World Spinus
species has a bright red face. In sum, those species that
remain at high latitudes during winter months when the
sun-angle is low have long-wavelength signals as expected,
and their southerly relatives lack them. I do not think
this informal check is a very accurate test of the prediction, but the result is consistent with the expectation
and so encourages further investigation.

altitude

The effect of altitude upon available light is not
the same as that of latitude, although there are some
similarities. As one goes up a mountain, the patch of
sunlight through the earth's atmosphere becomes shorter,
but the effects are not the same as those of moving toward
the equator. Variations in sun-angle merely alter the
length of the light-path through the entire atmosphere,
whereas elevation determines which layers of the atmosphere
light must traverse: high on a mountain sunlight simply
does not traverse the lower part of the atmosphere before
striking an animal. Since the atmosphere becomes denser

closer to the earth, relatively small changes in altitude
cause noticeable changes in ambient irradiance. Furthermore, the composition of the atmosphere differs at different heights above sea level, so the filtering effects
change.

The major changes in ambient light with increased
altitude appear to be higher irradiance levels and a subtle
shift toward a shorter-wavelength emphasis in light quality.
Figure 7-8 shows SOLREF simulations for the latitude of

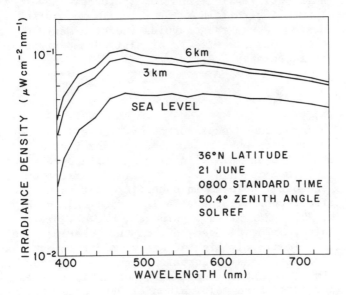

*Fig 7-8. Spectral irradiance depending upon altitude
above sea level (computer simulation).*

Death Valley, California at three elevations. The floor
of Death Valley is slightly below sea level (lowest place
on earth) and to the west lies Mt. Whitney at 14,495 feet
(highest place in the contiguous United States). The top
of this mountain is therefore about 4.4 km in elevation,
lying between the two upper curves in *fig 7-8*. From the
figure two effects of elevation may be noted. First, irradiance levels increase with altitude, the change being
less pronounced the higher one goes. And second, there
is a shift from an essentially flat spectrum to one that
peaks about 480 nm, that region of the spectrum we see as
blue. Whether these effects are quantitatively sufficient

to influence animal coloration remains an empirical question.

The consequences of these changes are not easy to
predict. One expectation might be that with short and
long wavelengths available for reflection, animals confined to high altitudes could evolve purple signals *(ch 5)*,
a color quite rare in most animals. Birds most evidently
confined to high elevations are the rosy finches *(Leuco-
sticte)* of the high Rocky Mountains. All three species
have pinkish-purple wing- and rump-patches. Nevertheless,
the most purple bird in North America is the varied bunting of Mexico, whose range avoids the high Mexican pla-
teau, so the possible effects of altitude on signal-coloration remain an open question.

general habitat

The foregoing discussion of available light for signaling disregards any influence of the habitat, and so
applies only to animals signaling in quite open areas. As
was pointed out in *ch 3*, absorption and reflection by
plants and other objects in specific habitats may have a
large effect upon light available for signaling *(e.g., fig
3-13*, p. 79). For present purposes, one may divide surface habitat into (a) open, direct-sunlit areas, and (b)
vegetated, indirectly lighted areas, and then ask what
differences in signal-coloration might be expected between
them. Following this, the underwater habitat is considered.
There are three major kinds of *open-area habitats*:
field and prairies, deserts, and the surface of bodies of
water. Naturally, what constitutes an open, directly sun-
lit habitat depends upon the size of the animal being considered: a short-grass prairie may be open to a deer but
heavily vegetated from a mouse's viewpoint. The major
light characteristics found in open areas are high levels
of irradiance and directness of the radiation. These are
the two requirements for effective use of signals that depend upon specular reflectance, dichromatism and iridescence *(ch 4)*, so one might expect such signal-coloration
to occur in these open habitats and not in vegetated ones.

The expectation is difficult to check, but iridescent
colors are particularly common in ducks, especially the
puddle-ducks (Anatinae), in hummingbirds (Trochilidae) and
in sunbirds (Nectariniidae). Ducks, of course, do frequent

open-water areas, and even the tree-nesting wood duck displays on open water, sometimes in bright sunlight. Hummingbirds live in vegetated areas, but males display their iridescent colors in bright sunlight by special aerial courtship flights before the female (Hamilton, 1965). On the other hand, many birds that live in open habitats have not evolved iridescent coloration, so there is no precise correlation. Open areas provide the opportunity for the evolution of iridescent coloration but do not demand it.

Vegetated habitats have the opposite characteristics with respect to ambient light: the absolute levels are low and the light is diffuse, due to filtration and reflection *(fig 3-14,* p. 80). Furthermore, the light quality is shifted toward a distinctly greenish illumination *(fig 3-13,* p. 79). Even in forested areas, however, some direct sunlight penetrates, often to ground level, so there is a continuum of lighting conditions between totally open areas and the deepest, darkest forests. D.H. Janzen *(pers. comm.)* has pointed out that near sunrise and sunset in stratified tropical forests a good deal of direct sunlight streams in nearly parallel with the ground, so that forest lighting is qualitatively different at different times of day. These problems make it difficult to check any expectations without quite detailed information concerning where and when animals actually display their coloration.

The expectations themselves are relatively straightforward: one expects highly reflecting signals such as white or yellow to provide maximum reflection in dark forest. Dark colors, such as violet, blue, green and even red, should be rare. A good check on this expectation is provided by Burtt's (1977) data on wood warblers (Parulidae). In general, warblers living in dense spruce and other northern forests have yellow or white signal-coloration, enhanced by contrasting black, whereas species displaying in more open areas have evolved other coloration. For example, the male blackburnian warbler is a treetop species that has a bright orange throat, the chestnut-sided warbler is a scrub species that has a greenish cap and chestnut sides and the redstarts *(Setophaga),* which often display in bright habitats, have red or orange. Nevertheless, the black-throated blue warbler is all-blue above and apparently lives in quite forested habitats, so simple correlations without specific knowledge of display-sites are only suggestive. More will be said about the coloration of wood warblers in a later section, when other

Table 7-II

Available Light and Signal Coloration

Optical Environment	Specific Circumstances	Characteristics of Illumination	Optimum Signal Characteristics
extremely dim illumination	night very deep forests deep ocean	very low irradiance	white reflectance bioluminescence
low sun-angle	sunrise sunset winter high latitudes	long wavelength low irradiance directional differences	white long-wavelength reflectance
high altitude	mountains	short and long wavelengths high irradiance	purple reflectance possible
open surface	fields & prairies deserts water surface	high irradiance directed light	specular reflectance dichromatism iridescence dark colors (red, blue)
vegetated above water	woods brush	low irradiance diffuse light greenish hue	white or yellow reflectance
under water	oceans lakes rivers	bluish hue diffuse light	high reflectance

factors can be taken into account.

As was noted in *ch 3*, two primary factors determine available light in *aquatic habitats*: depth in the water and reflection from the bottom (*figs 3-15* and *3-16*, pp. 82 and 83). Turbidity *(eq 3.14)* also plays a role in determining available light, but its effects as transmission

noise are possibly even more pronounced, as noted in a
later section. As depth in the water increases, the light
falls in level, shifts to shorter-wavelength emphasis and
becomes more uniform with respect to angle of view *(fig
3-15)*. These effects are somewhat offset by reflection
from the bottom *(fig 3-16)*. Unfortunately, consideration
of available light for reflection in aquatic habitats generates no clear expectations of signal-coloration, except
for the probable need of high reflectance. This need is
admirably met by the total reflectance produced by guanine
plates in fish scales (see *ch 4*), which may reflect all
ambient light or be modified as interference filters to
reflect selective colors.

   Table 7-II (opposite) summarizes some of the variations in light available for optical communication under
various environmental circumstances and the possible effects upon the variables of signal-coloration.

Optical Background

   Detection noise occurs when a receiver cannot distinguish an optical signal from its background. For convenience, backgrounds may be divided into four groups for discussion: those that are visually homogeneous (*e.g.*, sky),
those that have two types of homogeneity (*e.g.*, water's
surface), those that are regularly patterned (*e.g.*, cattail marsh) and those that are irregularly structured with
respect to optical properties (*e.g.*, coral reef). None of
these is really a distinct category, as emphasized in the
discussion to follow, but in order to identify relevant
environmental continua it is useful to begin with extreme
cases. Furthermore, the general ecological identification
of a species' habitat is only a first clue to the background of its optical signals: many animals choose specific display sites for a given kind of communication, so
detailed ethological studies are required to evaluate hypotheses proposed in the following sections.

   Although all aspects of signals are considered, the
sections below concentrate on colorational aspects because
these seem to be the aspects that presently yield predictive hypotheses. A powerful tool for analyzing background
effects on signals is the study of intraspecific variation,
but few examples are available. J.L. Brown (1963) found
that the crest of the Steller's jay was longer in Arizona
than in other parts of its range to the north and south.

The long crest thus correlates with open habitat having
greater visibility, but just why greater visibility might
lead to longer crests remains an unanswered question.
Such empirical findings are a useful adjunct to the approach taken in this section, and may lead ultimately to
generalizations that can be explained in terms of optical
principles, but I have not attempted to extract such generalizations from the literature. Rather, I have tried
to consider the optical factors of different backgrounds
and generate from these testable expectations concerning
characteristics of optical signals.

homogeneously bright backgrounds

There are at least four general kinds of optical
backgrounds that may be extremely homogeneous to the eye:
clear sky, snow, stretches of sand and open-water depths.
Of course, the visual homogeneity of sky may be broken
by clouds and a totally overcast sky may be almost homogeneous at the other extreme of weather. The strategy a-
dopted here is to scrutinize extremes of continua as a
first approach to how background may affect signal characteristics.

The homogeneous backgrounds have at least one over-
riding optical factor in common: they will almost always
be brighter than any object seen against them. In the
cases of clear sky and open-water depths, this factor occurs because the background is itself luminous, due to
scattering (ch 3); in the case of sand and snow the effect
is due to the high reflectivity of the substrate. There
are some probable exceptions to this generality. For example, a bird seen against the open sky away from the sun,
so that sunlight comes from behind the viewer, might be
as bright or brighter than the sky if the bird be very
highly reflecting. However, to have an appreciable effect
on the evolution of signal-coloration, this rare circum-
stance would have to be consistently the one in which the
animal communicates: a sufficiently unlikely possibility
that it may be disregarded for present purposes. A fish
near the surface of open water can reflect with its guanine-containing scales (ch 4) a specular flash toward a
viewer on a line toward the sun's azimuth--a flash that
would be brighter than the back-scattered illumination of
its visual background. This exception is more likely to

occur commonly than the exception against the sky; the effect decreases with depth as the illumination becomes more
diffuse and angularly homogeneous (ch 3). Finally, since
sand and similar homogeneous substrates do absorb some
light, any animal that absorbed less would become slightly
brighter than its background; again, there is a continuum
from highly reflecting, dry quartz sand to wet sand, mudflats and other homogneously dark backgrounds.

Because animals will usually be darker than their
homogeneous backgrounds, they are most conspicuous when
their coloration is dark. Although the backgrounds of
clear sky and water-depths are bluish in spectral emphasis,
their high luminosity often overwhelms the coloration.
Therefore, one does not expects animals to adopt signal-
coloration of the complementary color, unless that coloration could be quite dark. Figure 7-3 shows that the colors
complementary to blue are primarily yellow and orange. Although quantitative comparisons would have to be made,
taking into account the average luminosity of the sky or
water and particularly animal colorations, it seems likely
that a dark animal will nearly always be more conspicuous
due to its brightness contrast than would be a yellow or
orange animal due to its complementary coloration. Therefore, the expectation is that animals habitually signaling
against these homogeneously light backgrounds will show
primarily achromatic signal coloration: totally dark animals to emphasize shape, movement and orientational elements of behavioral signals, or else dark animals with internal white signal-patches.

It is difficult to check this expectation with regard
to sky because few animals are so aerial that most of their
optical communication must take place against the sky as
background. Of birds, only the swifts (Apopidae) are known
to fly high and virtually continuously during daylight
hours, so that any optical communication almost always must
be against the sky. So far as I can determine, the 60 or
so species of swifts in the world are all dark brown or
black, some with white markings on the rump, sides, throat
or breast. It is perhaps informative to note that the related crested swifts (Hemiprocnidae) have bluish gray plumages or browns with metallic gloss, with orange or other
color on the head, and unlike the true swifts do not spend
all of their time in the air.

One similarly expects open-water animals to have dark
signal-coloration: for fishes that are silvery colored as
concealing coloration, the expectation is for dark patches

used for open-water communication, as in schooling. In
general, this modest expectation is so overwhelmingly met
that only a few examples need be mentioned: menhaden with
the black spot behind the operculum, similar coloration
in the gizzard shad and alewife, and general silvery color
with smaller dark spots in other herrings and in the open-
water salmon species. However, without specific observations on how animals are signaling and under what precise
conditions, the coloration of fishes is even more difficult to evaluate than that of birds. Like birds, most
fishes engage in reproductive displays near a heterogeneous substrate where other principles of optical conspicuousness apply (see below), but unlike birds many fishes
change coloration in different communicative interactions.

Finally, the expectation of dark signals against light
sand is perhaps the most difficult one of all to check
unambiguously. Most beach species suffer predation, unlike
the high-flying, fast-wheeling swifts that virtually no
raptor can catch on the wing. Therefore, beach species
will be concealingly colored in general (see *fig 6-5*, p.
188, for an avian example), and one must have specific
information about their display behavior and morphology in
order to test the expectation. Many of the shorebirds
that frequent beaches in the winter in fact breed in heterogeneous environments of the arctic, so only winter plumage is relevant. In Florida during the winter the only
two species found consistently and commonly on the open
Atlantic beach are the black-bellied plover and the sanderling, a sandpiper. The plover has a black belly only in
summer, but its one obvious signal used during the winter
is the black wing-pits that show when the bird flies, and
hence may be a social signal for flight. The sanderling
similarly has a black outer wing and trailing edge with
a subterminal white stripe, along with a black mark longitudinally through its tail: all presumptive signals that
show when the sanderling flies.

Despite the difficulties in making meaningful preliminary checks, the available evidence supports the general
hypothesis: in the common homogeneous background situations,
the major factor governing conspicuous signal-coloration is
brightness contrast. Animals signaling with a background
such as sky, sand or water-depths tend to have achromatic
signal coloration: theirs is largely a world of black and
white.

bivalent homogeneous background

The surface of open water (*e.g.*, lakes and oceans) has special optical characteristics as a background because of the high specular reflectance *(ch 3)*. When viewed toward the sun, the water's surface may be extremely bright, like backgrounds discussed in the previous section; when viewed from other angles or on dim days, the appearance may vary almost to blackness. For this reason, I call the water surface a *bivalent homogeneous background*. Of course, it is not always homogeneous, either: a billowy water surface may have irregular wavefronts, white-caps and other visual heterogeneities. Again, the strategy is to take the extreme case of homogeneity and see what effects it suggests.

Figure 7-9 diagrams the bivalency of reflection of relatively calm water. An observer to the right of the

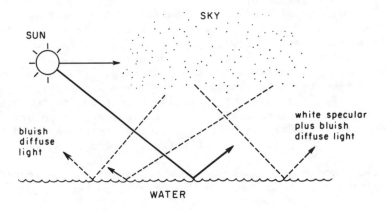

Fig 7-9. Reflection off water toward and away from the sun differs in total radiance and spectral distribution. An observer to the right of the diagram sees an emphasis of bright, white, specular reflectance, whereas an observer at the left sees dimmer, bluer reflectance from the diffuse skylight.

diagram sees the surface as very bright and achromatic, the specularly reflected sunlight overwhelming the component from skylight. This situation predicts essentially

the same characteristics as the homogeneously bright backgrounds of the previous section: largely dark animals, or
dark with white signal-patches. An observer to the left
of the diagram, however, sees the water's surface as being
primarily bluish and not nearly as bright. When conditions
are overcast, the water will not look so bluish, but it
will be yet darker, and directional differences will be
less critical than under the clear conditions diagramed
in *fig 7-9*. Such conditions of viewing or illumination
lead to expectations concerning signal-coloration quite
different from those of the homogeneously bright background.

An animal on the water's surface as seen by an observer on the left of *fig 7-9* will be strongly illuminated by the sun's direct rays under clear conditions *(table
7-II*, p. 218), making iridescence and chromatic coloration
possible. Furthermore, with the background darker than
other homogeneous backgrounds considered above, it is now
possible to achieve contrast by making a signal lighter
than its background.

Because of the bluish emphasis of the background, if
the signal is to be chromatic, it will contrast best if it
is a complement to blue, which from *fig 7-3* (p. 203) can
be seen to be green, yellow or orange. Yellow and orange
biochromes each reflect a broad spectral band *(ch 4)* and
have peak reflectances near the maximum sensitivity of the
eye *(ch 5)*: they are intrinsically bright colors for these
reasons, and hence would make excellent signal-colors under these conditions. Green reflects a narrower band and
hence would appear somewhat darker, unless mixed with white;
in other words, under these special conditions a *de*saturated green might appear more contrasting than a saturated
color of the same dominant wavelength, an exception to the
usual rule *(table 7-I)*.

How can one then design the ideal animal for signaling
under the bivalent conditions of the water's surface as a
background? Basically, the best compromise appears to be
a primarily black or dark animal with light signal-patches
that are either (a) iridescent, (b) saturated yellow or
orange, (c) desaturated green or (d) white. It is also
possible that the reversed polarity would be almost as contrasting: light body, probably white, with dark markings.

The expectation concerning signal-coloration is difficult to check. The primary animals that must signal a-
gainst a background of the water's surface are pelagic
birds that spend most of the time on the water itself ra-

ther than on the wing. Furthermore, only the winter plumage of such birds can be considered because during summer months the birds are all nesting somewhere on land,
where a different optical background occurs. Lastly, the
birds need to be social in their winter plumage, either
for integration of large flocks or for winter courtship
on the water's surface. Bay ducks and sea ducks (Aythy-
inae) seem to meet the conditions for a check.

Among the bay ducks, species of the genus *Aythya* all
fit the expectations more or less rigorously. The redhead
and canvasback males have russet-orange heads, black breasts,
black posterior coloration and white bellies. The male
ring-necked and tufted ducks, and both species of scaups,
have iridescent purple or greenish heads, black or iridescent fronts, white on the sides and belly, and dark
posteriors. Females of all these species are dull brown
with white markings on the face. The male *Bucephala* ducks
(bufflehead and goldeneyes) are white beneath with dark
heads having white markings and their females tend to have
brownish-orange heads. All the males also show iridescence on the heads.

The sea ducks are differently patterned from the bay
ducks, but still agree with the expectations, with the exception of the harlequin duck, mentioned below. The scoters
(Melanitta) and oldsquaw males are black and white, the
former being primarily black with white markings, the latter
reversed (in winter). Females are dark brown and white.
Furthermore, males of all three species of scoters have
orange patches about the bill. The eiders *(Somateria)* all
have black and white males with dark brown females, but the
males also possess chromatic patches. The Steller's eider
has a russet-orange breast and belly and the king eider has
an orange bill and frontal shield. The common and spectacled eiders have desaturated green about the head--a most
unusual avian color. There are, however, two exceptional
examples not expected from the considerations of the background. The male king eider has pale bluish feathers on
the crown and back of the head, for which I have no *post
hoc* explanation, and the male harlequin duck lives up to
its name in a riot of color patches in black, white, blue
and orange (the female is dark brown with white head-markings).

I was so stuck by the degree to which the expectations
correctly predicted coloration of bay and sea ducks that
I searched for a taxonomically distinct group on which the

same check could be made. The only other groups meeting
the criteria for a check appear to be the alcids (Alcidae)
and penguins (Spheniscidae), polar oceanic birds that are
essentially unrelated ecological equivalents. As far as
I can determine from handbooks, all species of both groups
are black and white in winter. When species possess chromatic patches, as in bill color, the coloration is always
yellow or orange, although iridescent and desaturated
green colorations have apparently not evolved in either
group.

regularly patterned background

   One type of heterogeneous background is a regularly
patterned one with repeating elements of some kind. There
are many such regularly patterned backgrounds created by
man's agricultural efforts, but these are so recently created in evolutionary time that it is unreasonable to expect animal signals to be adapted to them, even in species
that now live in man's crops. Probably the most common
regularly patterned natural backgrounds consist of vertically oriented, sparsely branching plants that provide a
visual background of a repetitively barred array. Some
habitats that come close to this ideal are marshes, natural prairies and beds of eel grass.
   There seem to be at least two possible strategies
for conspicuousness against a regularly barred pattern:
blocks of uniform coloration or regularly barred patterns
of opposite orientation. I have not found an adequate way
to make an informal comparative check of these expectations,
so merely cite a few examples that might illustrate the
points. One of the commonest grassland prairie birds
whose coloration is not clearly concealing is the male lark
bunting: solid metallic black with large white patches in
the wings. (The female is striped brown like most emberizine sparrows.) A bird typical of and confined almost entirely to homogeneous cattail marshes is the American bittern. This species is striped longitudinally, and when
assuming its "freezing" posture with bill pointed toward
the sky is extremely concealing against its background.
When displaying, however, the bird is often oriented horizontally so that its striped pattern is orthogonal to that
of the background; and when flying, the large black patches
in the wings are quite conspicuous. It is difficult to i-

dentify coloration of fishes habitually living in eelgrass
beds because coloration is so changeable in fishes; for
example, the kelpfish is cryptically brown when in kelp
and green when in eelgrass. If fishes were cryptically
vertically barred, they could become conspicuous either
by aligning their bodies vertically to make the barring
oriented at right angles to the background (the bittern's
strategy) or else by undergoing color-changes that bring
on longitudinal stripes.

With homogeneous backgrounds considered previously,
virtually any movement of the animal is conspicuous. In
a regularly patterned environment, however, certain movements may be more conspicuous than others. For example,
in the marsh habitat of the green heron, movement of the
vertical vegetation by wind is primarily horizontal, and
many of the heron's display movements tend to be vertical
(Meyerriecks, 1960).

other backgrounds

Most animals communicate in a variety of ecological
situations or in habitats where the visual background is
quite heterogeneous. In such cases, it may require considerable study to determine the common optical backgrounds
for display, as one needs to know how the receiver sees
the sender in order to specify the background against
which the sender is perceived. In many cases, it may turn
out that although the habitat in general is quite heterogeneous, certain fixed display sites are used that render
only one or a few kinds of optical background common. For
example, several of the mimic thrushes (Mimidae) sing primarily from high, exposed sites such as tree tops. The
female is usually lower in the vegetation and hence often
sees the male singing against the open sky. In species
where evident optical displays accompany singing--as in
the jump-displays of the mockingbird (*pers. obs.* and Baylis, *in prep.*)--one expects the coloration to follow the
rules for homogeneously bright backgrounds (above). In
the case of the mockingbird, the rules hold because the
species is primarily dark gray and white. The related cat-
bird, however, commonly sings from undergrowth, so it is
not surprising to find chromatic coloration on this species (conspicuous russet undertail coverts).

It is difficult to deduce any expectations for signal
characteristics when the optical background varies a great

deal. One might expect simply that coloration of a forest bird or coral-reef fish would have various elements of pattern and coloration, some of which would be more conspicuous against some backgrounds and other of which would be optimumly conspicuous against other backgrounds. One arrives, therefore, at a vague expectation of complex coloration in such animals--hardly a testable or very useful expectation.

At best, one can articulate research tactics for uncovering possible influences of the background on signal characteristics. The minimum effort would involve a frequency distribution of backgrounds against which the intended receiver sees the sender under natural conditions, optical characterization of the backgrounds, and then perusal of *table 7-I* to find out how general principles of conspicuousness might be manifested as specific display characteristics. For example, Burtt (1977) investigated the coloration of wingbars and tailspots in wood warblers (Parulidae) by measuring the ambient irradiance in coniferous and broadleaved forests, the reflectances of leaves in those forests (general background coloration), and the frequency of occurrences of species in the two forest-types. He found that white is the most contrasting color, based on a three-dimensional model of surface-color space (to be shown later in this chapter). However, in broadleaved forests, yellow and orange are nearly as conspicuous as white. His censuses show that all the coniferous warbler species have white signals, as expected, and that most of the broadleaved warblers are similarly colored, but a few have yellow or orange signals. Despite the many sources of error in such studies, Burtt's example shows that it is possible to begin making sense of signal-coloration in complex habitats by means of detailed study and comparative correlations.

A different approach that seems worth pursuing turns around the independent and dependent variables of investigation. In this monograph, my approach has been to specify optical principles, search for ecologically relevant situations, and then predict animal signal characteristics. Another approach would be to find consistent correlations between ecological situations and animal characteristics, and from these generate new hypotheses about the optical principles governing the situation. For example, one specialized cluster of ecological niches in birds relates to finding insect prey beneath the bark of trees. By surveying the coloration of unrelated bark-creeping species,

it might be possible to find certain elements of coloration in common, and from these begin to speculate on
optical principles governing the correlation. The striking morphological fact about woodpeckers (Picidae), for
example, is that they are primarily black and white; in
most species the male also has some red coloration somewhere, often on the head. The exception among North American woodpeckers is the common flicker, a primarily brown
species with yellow or orange in addition to red, white
and black. This species commonly feeds upon ants on the
ground, and hence is an exception that tests the rule.
The black-and-white warbler also feeds on treetrunks and
has coloration very similar to that of woodpeckers, whereas most other wood warblers are very differently colored.
Nuthatches (Sittidae) are also black and white, although
these trunk-creeping birds have bluish dorsal coloration
and some species have other coloration as well (brown-
headed nuthatch and red-breasted nuthatch). I have no
optical explanation for this convergence of coloration in
trunk-creeping birds, but note that the ecologically similar brown creeper has no conspicuous black and white plumage, but rather is concealingly counter-shaded with a
brown dorsum and lighter underparts. Furthermore, the
blackpoll warbler male is visually similar to the black-
and-white warbler, yet does not commonly engage in trunk-
creeping, so the correlations are by no means perfect.
Yet there is sufficient association of an ecological situation with presumptive signal-coloration that one feels
detailed research might prove promising.

The few principles extracted from considerations of
backgrounds are summarized in *Table 7-III* on the next page.
Hopefully, further research will greatly enlarge this
table.

Transmission

Any physical alteration of the signal sent that causes
equivocation in the receiver is transmission noise. In optical communication, it is useful to distinguish *translucency* of the medium, in which small suspended particles
may disrupt the signal, from *opacity*, in which large objects partially obscure view of the signal-object.

Table 7-III

Optical Background and Signal Coloration

Optical Background	Specific Circumstances	Optical Characteristics	Optimum Signal Characteristics
homogeneously bright	clear sky open sand water depths	high radiance	black with white patches black patches
bivalent homogeneity	water surface	high radiance OR low radiance bluish hue	black with: yellow orange light green iridescence
regularly patterned	cattail marshes prairies eelgrass	vertical barring	uniform coloration horizontal striping

translucency

Noise in a translucent medium is due to small particles suspended in the medium between the sender and the receiver. For convenience of reference I shall call this "translucency noise." Its principal effects are to blur the sharpness of visual images, reduce the amount of reflected light reaching the eye, and selectively absorb or scatter spectral components of the transmitted signal.

Two principal types of translucency noise occur. In terrestrial environments, water vapor in the form of mist, fog, *etc.* is the primary source, whereas in aquatic environments turbidity due to suspended particulate matter is the important source. Their effects are similar, except in spectral transmission. Water vapor appears to cause primarily Mie scattering *(eq 3.10)*, the spectral effects of which are complicated, whereas absorption due to at least some kinds of turbidity increases with spectral frequency *(eq 3.13)*.

There appears to be no easy way to combat the first two effects of translucency noise (blurring and low light

level) except by decreasing the distance between sender
and receiver when the noise occurs. One might expect optical signals in translucent media to consist of large
patches of color and rather gross movements to combat the
loss of image-sharpness, but I can find no critical evidence bearing on this hypothesis. Reduction in the total
amount of light received is probably not serious compared
with the loss of image, and in the extreme case both factors cause total opacity. The solution to this problem is
simply to use another channel. Wootton (1971) notes that
a stickleback living in tea-colored water has less well-
developed optical signals than its congeners living in
transparent media.

Animals should be able to combat spectral effects of
translucent noise by reflecting those wavelengths that
penetrate best. When the noise is due in part to scattering rather than simple absorption, then the back-scattering may create an optical background of the opposite spectral extreme and thus increase the signal-to-noise ratio.
Baylis (1974a) notes that the yellow signal-coloration of
a cichlid fish may be an adaptation for getting the signal through its turbid environment. G.W. Barlow (1974)
notes that cichlid species living in turbid waters generally have yellow, orange or red markings (especially
about the eyes), whereas those living in clear waters have
blue and green as characteristic coloration.

Spectral effects of terrestrial water vapor require
more study in relation to animal signals. Depending upon
the size-composition of droplets in various kinds of fog
and mist, the Mie scattering could create various spectral effects *(ch 3)*. Fog lights on automobiles are ordinarily of long wavelength, possibly an empirical choice
based on what light penetrates best and back-scatters
least toward the driver. Back-scattering is not as important in Mie scattering as in Rayleigh scattering, and in
any case an animal sender communicating by reflected light
is primarily concerned with penetration to the receiver.
It could be that the signal-coloration of marsh birds like
the yellow-headed blackbird and red-winged blackbird shows
long-wavelength reflectance to penetrate early morning
mist. Minnaert (1954: 257) states that the sun is usually
white when seen through fog or mist, but may be red--an
occurrence he attributes to Rayleigh scattering by very
small droplets. Foglights and animal signals may therefore be adaptations to rare conditions of Rayleigh scat-

tering, as well as more usual conditions of Mie scattering.
Empirical studies in specific signaling habitats and conditions would be most useful.

opacity

     Opacity noise is due to any physical object that part-
ly blocks the receiver's view of the signal-object. The
principal source of opacity noise of transmission in both
terrestrial and aquatic habitats is plant material. The
chief effect of the noise is to obscure parts or all of
the signal-object, and the obvious way to combat this noise
is to avoid it by communicating in a clear environment, or
at least at a short distance. Thus, many songbirds perch
on the outer branches of trees or at the tops of trees, or
fly while displaying; many mammals seek the crest of a hill
or the top of a boulder for optical communication, *etc.*
     Given that an animal lives in and must display in a
vegetated environment, how could its optical signals be
designed to combat opacity noise? Apparently this question
has been given little attention, so it is not even clear
as yet how it should be approached. In *ch 3* I suggested
that a beginning might be made by modeling the simple case
of circular "holes" in the vegetation through which circular signals are viewed. Considering the case of a single
"hole" in the vegetation, the receiver can view the greatest part of the signal by minimizing the distance to the
hole and maximizing the distance to the signal *(eq 3.17)*.
This beginning appears to predict a possible spatial arrangement for optical communication in small forest songbirds. MacArthur (1958) noted that in some species of
boreal wood warblers the female forages low whereas the
male displays from the tops of trees. This spatial arrangement often is such as to combat opacity noise, as shown in
fig 7-10.
     Extension of the kind of model represented by *eq (3.17)*
would have to deal with at least four complicating factors.
First, the size of the "viewing holes" in the environment,
their geometric pattern in the plane normal to the line of
view, and their spacing in that plane all affect the probability that a signal-object of a given size will be seen
in its entirety. Second, the number of "planes with holes"
between the sender and receiver, as well as their distances
from the communicating animals, must be taken into account.
Third, the shape of the "holes" relative to the shape of
the signal-object must be considered. And last, one must

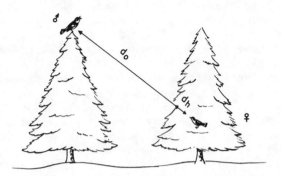

Fig 7-10. An optimum strategy for observation when opaque objects are in the line of sight is to minimize the observer-to-object distance (d_h) and maximize the object-to-sender distance (d_o). See eq (3.17), p. 86.

deal with the problem of whether or not is is necessary for the receiver to see the entire signal-object. This decision depends upon the nature of the information-carrying sign vehicle (see Johnston, 1976, for a theoretical discussion). It is premature to consider all these ramifications here, but a few comments relative to them can be added.

One need not see an entire signal-object in order to recognize it. Often, the view of part or several disconnected parts of the object is sufficient to construct the object perceptually. One strategy of animal signaling, then, may be to create rather large, homogeneous signals that allow visual reconstruction when only parts of them are seen. The opposing strategy is to create a small signal, the entirety of which is viewed.

In order to test comparatively the hypothesis of opposing strategies one requires predictions of factors that favor the strategies differentially. Body-size of the sender is one such factor. If a woodland bird is small, it cannot use the strategy of "perceptual reconstruction ." even if the entire bird were homogeneously colored with bright signal coloration, not enough of it could be seen for reliable reconstruction. Therefore, one may expect small woodland birds to use small signal-patches of coloration that may be seen in their entirety, whereas large

woodland birds may be homogeneously colored for signaling.
However, once again this depends upon the nature of the
sign vehicle involved. The expectation does, though,
appear to have some merit. Male wood warblers are rarely
homogeneously colored, having instead small patches of presumed signal-coloration on various parts of their bodies
(caps, facial marks, breast-bands, *etc*.). Large woodland
birds, however, are often homogeneously colored, as in
many jays and the blue-gray tanager (blue color), summer
tanager and cardinal (red), and so on. Other large woodland birds are boldly marked in only two colors, such as
many species of orioles (orange and black), scarlet tanager (red and black), evening grosbeak (yellow and black),
etc. There may be other reasons for this correlation between body-size and coloration, however, such as differences in mean signaling distance.

More attention needs to be paid to the systematic
optical inhomogeneities of environments relative to opacity
noise. For example, the branches of coniferous trees tend
to radiate horizontally from a single trunk, whereas
branches of broadleaved trees tend to grow upward at an
oblique angle and then divide into smaller branches with
irregular angular orientations. In essence, a coniferous
forest approaches a layered optical environment that may
allow better viewing in the horizontal direction than would
a broadleaved forest. Another woodland example of structual inhomogeneity is caused by the fact that leaves are
in favorable places for absorbing light. One can often
see well vertically near the trunk of a tree, but not toward its periphery. A related inhomogeneity is due to
the lowest branches being the oldest, and hence the thick-
est. By reasoning similar to that behind the development
of *eq (3.17)*, one might expect that it is easier to view
something from below than from above. A squirrel hunter
told me that a strategy for detecting his prey is to lie
on the ground looking up through the trees, and this strategy may be another factor favoring the differential heights
of wood warbler sexes *(fig 7-10)*.

Similar optical inhomogeneities occur in environments
other than forests. For example, marsh vegetation tends
to be vertical arrays of long-stemmed grasses and sedges.
This array causing opacity noise might, again, be combat-
ted by different strategies: either by use of long, thin,
vertically aligned signals that could be viewed entirely;
or by wide, horizontally aligned signals that would have
to be perceptually reconstructed. In this case, the prob-

Table 7-IV

Optical Transmission and Signal Characteristics

Optical Transmission	Specific circumstances	Optical Characteristics	Signal Characteristics
translucency	suspended particulates	blur images low radiance change spectrum	short distance large signals gross movements
	a) mist, fog, *etc.*	(spectrum unsure)	(see discussion in text)
	b) turbidity	long-wavelength transmission	red, orange, yellow
opacity	plants rocks coral	obscure signal-object	various spatial arrangements very small OR very large signal-patches

---

blem with the first strategy is that the signal might be mistaken for its background (part of the problem of detection noise, considered previously), as when the bittern "freezes" vertically as a deceptive posture to escape detection. Aquatic environments also have optical inhomogeneities, such as kelp beds, layering of coral heads and so forth.

In sum, one can state that opacity noise of transmission is an obviously important consideration of both the spatial arrangements of communicants and the physical structure of optical signals. However, it is just as obviously a complex matter that requires considerable attention before hypotheses can be generated from knowledge of the environment that predict design characteristics of optical communication.

Table 7-IV summarizes some aspects of signal-transmission and their possible effects on optical communication. It represents a bare beginning of an area that hopefully will receive more experimental attention.

Concealment and Conspicuousness

This chapter has emphasized characteristics of presumed social signals that render them visually conspicuous, ignoring the needs for concealment dealt with in *ch 6*. In this section, I explore how these opposing needs interact to create compromises in behavior and morphology.

I shall take the extreme case in which the sensory capacities of the undesired observers (*e.g.*, predators) and the desired observers (*e.g.*, social companions) are roughly similar. This condition will not always be true, but in general small fishes suffer predation from larger piscivorous fishes, songbirds are preyed upon by raptors, and so on. Furthermore, under conditions of similar sensory capacities the dilemma of how to be both concealed and conspicuous is heightened, so that investigating the extremes may lead to general solutions.

Under these conditions of similar sensory capacities, the sender's only recourse is to effect some kind of difference in viewing situations for undesired and desired observers. There are at least three interacting variables in viewing situations: time, distance and orientation. Orientation may be further decomposed into orientation with respect to background, specific addressees and the part of the sender observed. These viewing situations cannot easily be separated, but it is possible to consider them separately to see how optical signals might be adapted to them.

timing

The obvious solution to the concealment-*vs*-conspicuousness problem is simply to be concealed generally and conspicuous only for brief periods of communication. This strategy is used so generally among animals that one may not stop to consider that optimum communication in the absence of opposing needs might dictate that animals be habitually conspicuous. Ernst Mayr *(pers. comm.)* pointed out to me that birds on South Pacific islands, having relatively fewer predators than on continental areas, are generally more brightly colored. Furthermore, larger North American songbirds—such as orioles *(Icterus)*, tanagers *(Piranga)* and various large finches (*e.g.*, cardinal, rose-breasted grosbeak, evening grosbeak, blue grosbeak)—may be less subject to predation than smaller songbirds, and the males of all these species are quite brightly colored. Smaller songbirds, presumably suffering greater predation, are more concealingly colored (*e.g.*, emberizine sparrows, tyrannid flycatchers, parid tits). And when the needs of optical communication dictate bright coloration in male songbirds, as in the parulid warblers, the birds molt back into concealing plumage after the breeding season, unlike larger songbirds that retain their bright coloration the year-round.

In general, then, an animal that communicates optically will probably be as conspicuous as predation pressure allows. In each case, some balance must be struck between failure of reproduction because of the inability to attract a mate or hold a territory on the one hand, and failure of reproduction due to death or injury by predation on the other. Animals solve this dilemma by timing conspicuousness in two general ways: either by color-changes or by display of normally hidden conspicuous coloration. Many male birds change into special conspicuous coloration for the breeding season, molting back into concealing plumage thereafter, whereas many fishes, some cephalopods and a few other animals can change color over shorter periods of time *(ch 4)* to assume conspicuous coloration for communicational purposes. The fact that animals have hidden coloration displayed only for communication is so well known (*e.g.*, Tinbergen, 1951) that one needs only to be reminded of rump and tail colors of mammals; wing-stripes, tail-spots and rump-patches of birds, *etc.*

We ordinarily think of display coloration in terms of a hidden color-patch that is revealed by movement, but B.D. Sustare and E.H. Burtt *(pers. comm.)* have suggested that something like the reverse is also possible. Some animal movements are so rapid that visually conspicuous coloration at rest could become concealing in motion. For example, the wingbeat frequencies of hummingbirds and some insects exceed the critical fusion frequency *(ch 5)*, at least at moderate illuminance levels, so that a bold pattern of black-and-white might fade to a more concealing gray in flight. It is also possible that rapid alternation of bright colors could lead to a perceptually fused color that is concealing. However, I know of no cogent examples of this theoretically possible strategy.

T. Johnston *(pers. comm.)* points out that in many situations predators can more easily capture stationary prey. Therefore, one may expect stronger selection for

concealment in the resting postures. Furthermore, motionlessness *per se* is concealing *(ch 6)*, so in general one
expects more optical signaling in moving than in stationary situations.

background

Without changing coloration in any way an animal can
be concealed against one background and conspicuous against
another. This solution to the opposing needs is probably
utilized to a much greater extent than has been noted in
the literature, although Moynihan (1975) makes a specific
mention of this principle in relation to cephalopod signals. For example, the brown thrasher is generally colored
concealingly, being counter-shaded brown above with dark
streaks on a light belly below. On the ground, the thrasher is quite inconspicuous, but the male almost always sings
from the very top of a tree, where his mate and neighbor-
ing males see him as a dark silhouette against the open
sky. Even in this conspicuous position, a hawk from a-
bove will not see the thrasher as clearly as will conspecifics because the optical backgrounds differ. Therefore,
two principles concerning background may operate in animal
communication: choice of a different background for display than for other activities, and choice of a site in
which the background is different for social observers and
potential predators.
    There is an interesting example among fishes that
interrelates background, body shape and coloration. As
noted previously, two principles of optical conspicuousness are brightness contrast and spatial repetition *(table 7-I)*. Vertically barred and horizontally striped
fishes are equally conspicuous in this regard, so one may
ask what factors dictate the pattern to be assumed by a
particular species. The answer, most probably, relates
to the backgrounds that make the patterns concealing under non-display circumstances. A long-bodied fish that
swims rapidly in open water will likely be tracked readily by predators, and so is longitudinally striped as a
mechanism of motion-deception *(ch 6)*. Yet stationary, it
is conspicuous due to factors mentioned above. Conversely, a deep-bodied fish adapted for twisting slowly through
vegetation will be concealingly colored with vertical bars
by the principle of matching coloration *(ch 6)*. Display-

ing in the open, moving rapidly through vegetation to create a flicker effect (ch 5), or orienting its body vertically all render the fish conspicuous. For example,
the slender remora, an open-water fish associated with
sharks, has black and white horizontal stripes, whereas
the laterally flattened spadefish has vertical barring
(fig 7-5, p. 207). The most convincing cases come within
individual species, as among certain wrasses (Halichoeres)
and parrotfishes (Scarus), where the young animals or females (or both) are longer and thinner than the males,
and also tend to be clearly horizontally striped, in contrast to the uniform coloration or indistinct vertical
barring of the males.

Exceptions test the rules, and it would be instructive to know why the pelagic pilotfish (a slender jack that
follows sharks) is vertically barred. Magnuson and Gooding (1971) report that a pilotfish apparently defends a
shark as a moving territory, so the conspicuous coloration
might be involved in intraspecific aggressive display.
However, when chasing another pilotfish, the chaser usually
blanches to a subtly counter-shaded, much less conspicuous
coloration. The authors suggest that the barred coloration might be aposematic: coloration to warn their potential predators away. Questions still remain, however, because the remora, which also accompanies sharks, is also
black and white, but has the expected longitudinal stripes
of a pelagic fish (fig 7-5). The example illustrates how
difficult it may be to account for the exact display coloration of an animal, even when something is known of its
habitat and behavior.

addressee-specificity

In special cases, it is possible that the sender can
orient so that it is conspicuous to a particular addressee
and less so to other observers, even against the same kind
of background. This phenomenon has received little attention and probably depends on various optical mechanisms,
so a few examples seem worth exploring. In some cases,
addressing a signal is a fairly straightforward matter not
involving any special optical principles. For example, a
displaying peacock orients his raised tail with the colorful surface facing the female. From behind, the coloration
is noticeably duller and more concealing, and because the

tail is planar, from the side it presents a small image. In other cases, addressee-specificity may be more involved.

In *ch 6* the ability of fishes to reflect a bright pulse of light from their guanine-containing scales was noted in the context of startle-deception. The fish is momentarily conspicuous, but immediately becomes concealed by swimming in a new direction that may minimize the body area seen by the potential predator. An analogous optical ploy is u- tilized for social communication, in which the display of bright specular reflectance is followed by lateral or other- wise conspicuous orientation of the sender. Male fishes of many species, for example, perform such a display be- fore females at a possible spawning site. It is not to- tally clear how fish control the specular flash of light, but it seems likely from considering the nature of their scales that orientation perpendicular to the sender-receiver axis could play a role. It may be that in usual postures the orientation of each scale is a little different, pro- moting no overall specular flash, and that special body postures that align the planes of many scales create the conspicuous signal.

A somewhat analogous case may occur in the male mal- lard, whose head appears green, purple or black in various situations. The exact basis of the coloration has not been scrutinized (A. Brush, *pers. comm.*): it could be dichromatic but most likely is an interference phenomenon *(ch 4)*. In any case, I have observed that viewing the head feathers of specimens normal to the feather-plane maximizes the green coloration, whereas viewing them end-on down the shaft maximizes the purple coloration. In most diffuse light the feathers appear nearly black. Mallards change their head colors during display by piloerection, and per- haps also by specific orientation with respect to sunlight. It is evident, for example, that in the "head-round" court- ship display posture (Lorenz, 1941; Weidman, 1956) the feathers are erected and the head appears purple to the viewing female. During other displays, such as the "grunt- whistle" the feathers appear to be depressed, flashing green coloration toward the female. In the display known as "showing the back of the head to the female" the male leads the female, turning his head from side to side, and hence possibly flashing purple and green alternately. (It is difficult to observe this display from the female's viewpoint, so I have few critical notes as yet.) The point is that the male mallard is able to address conspic-

uous coloration toward the intended receiver, and to some
extent appear less conspicuously colored to other observers.

Finally, there is the phenomenon discovered by Hamilton (1965) in which the male Anna's hummingbird orients
his courtship display with respect to the sun's rays and
to the female, so that she sees his iridescent cap, throat
and back. To observers at other angles the bird may simply look dark.

general orientation

The orientation of the sender with respect to an intended conspecific receiver can help to separate concealment and conspicuousness without specific recourse to different backgrounds or different colorations of the same
body part, as discussed in the foregoing sections. In
those cases, the undesired and desired observers see the
same body parts of the sender, but perceive them differently because of background or special reflection. A more
general phenomenon of display-orientation is analogous to
display-timing (above): the sender orients so that the intended receiver simply sees some different part of the
sender's body than does the general observer. Like timing,
this principle of orientation is so well documented (*e.g.*,
Tinbergen, 1951) that one needs do little more than point
it out for sake of completeness. What does seem worth
exploring is that these orientational aspects of signaling
predict certain accompanying adaptations of morphology,
particularly shape and coloration.

An excellent example of an adaptation of shape is the
display dewlap of certain small lizards. A male orients
laterally with respect to his intended receiver, and lowers
the brightly colored dewlap. Seen from the side, the position of another male or female to which he is displaying,
the male's signal is quite conspicuous. However, because
the dewlap is very thin, the male viewed from above, from
in front, or from behind shows no bright coloration and
remains concealed. This example is closely related to
that of the peacock mentioned previously, except that the
bird specifically orients his tail *per se* toward the female: the signal is addressed by moving specific body parts.
There is therefore a probable continuum between kinds of
orientations that render a signal conspicuous to some ob-

servers and not to others.

Burtt (1977) attempted to see if the expected correlation between coloration and use (concealment or conspicuousness) could be documented quantitatively in wood
warblers (Parulidae). He characterized the coloration of
warblers and their optical backgrounds by means of a three-
dimensional color-space having axes of dominant-wavelength,
excitation purity and relative luminance (*table 5-I*, p.
138). The exact scaling considerations are too detailed
to recount here; see Burtt (1977). He then measured the
reflectances of leaves and warbler colorations spectro-
photometrically and also measured spectroradiometrically
irradiances in various forest habitats where warblers occur. From the spectral products of these two elements,
he calculated the average radiance of various warbler colors and various leaf colors, and then plotted all in the
three-dimensional color-space, as shown in *fig 7-11*. The

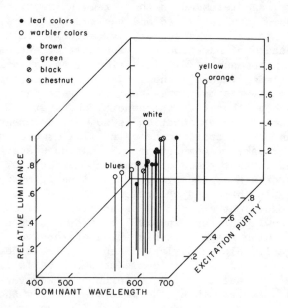

*Fig 7-11. A three-dimensional space in which each leaf
and warbler color plots as a point under specified conditions of illumination. In this example, some plumage colors
plot near leaf colors, whereas others (blues, white, yellow
and orange) plot farther from the leaves, and hence are
more conspicuous. (After Burtt, 1977.)*

diagram shows that warbler coloration of blues, white,
yellow and orange plot apart from leaf colors, whereas
warbler coloration of green, browns, gray, black and
chestnut plot among the leaf colors. One therefore expects these latter colors to be large areas of dark dorsal coloration for counter-shading and matching coloration
(the concealing part of coloration, *ch 6*), whereas the
blue, white, yellow and orange colors on warblers should
be smaller areas used for optical signaling. With a few
exceptions, this is just the pattern found in nearly 50
species of warblers. The result goes only one step beyond intuition, but does illustrate a method for objectively assessing which colors should be concealing and
which conspicuous, and hence predicts to some extent the
spatial arrangement of these colors on the animals.

distance

Because an animal is usually detected visually by
predators from afar and by conspecifics from nearby, it
would be potentially useful to be concealed from afar
while being simultaneously conspicuous nearby. There appear to be at least three ways in which this goal could be
accomplished with special coloration.
First, an animal could be colored so as to engage in
the deception I call element-matching *(ch 6)* when seen
from afar, but not when seen closeby. I was led to this
hypothesis by watching mountain bluebirds in western North
America. The male is often described as "sky-blue" (*e.g.*,
Robbins *et al.*, 1966: 234), but he obviously does not
blend into the sky when seen against it. However, this
species frequents open high meadows, where it perches in
trees that, when seen against the sky, have irregular
patches of blue sky seen through the green foliage. At
least under certain conditions of viewing, male mountain
bluebirds perched in such trees are difficult to detect
because they appear as just another patch of blue in the
foliage. An exact match in brightness with the sky is
unnecessary because the tree is such a mosaic of dark and
light patches that brightness-contrast is less important
in conspicuousness than is contrast in hue. Directly a-
gainst the open blue sky the male looks noticeably dark
and contrasting, of course. Furthermore, up close the
male appears very contrasting in both color and bright-

ness with the darker green foliage, and its avian shape
becomes evident so that the bird is highly conspicuous.
The male mountain bluebird is in fact conspicuous nearby
and at least sometimes highly concealed at a distance.

Another way in which an animal could be concealed
from afar but conspicuous nearby depends upon spatial fusion of color *(ch 5)*. An array of bright colors perceptually fuses at a distance to achromatic, desaturated coloration, given the proper selection of colors. Possible examples of this strategy occur in various parrotfishes *(Scarus* and *Sparisoma)*, where each scale may be edged in a col-
or that contrasts with the general ground color, or where
some scales are one color and some are a different color.
These fishes never appear brightly colored to me from a
distance, but when I dive close to them, they appear
brightly colored and highly conspicuous. Some quail *(Cal-
lipepla* and *Lophortyx)* show ventral scaly coloration with
must contrast nearby that blurs to a general gray in the
distance. The yellow warbler has bright red streaks on
its breast that become virtually invisible at several me-
ters distance.

A third mechanism relating to distance concerns conspicuousness of barred or striped patterns. If the contrast between dark and light stripes is not too great,
the pattern will be detected only at some optimum distance
(ch 5). Furthermore, since contrast-sensitivity optima
differ in spatial frequency among species *(fig 5-15*, p.
144), a barred animal could be conspicuous to one species
(say, its conspecifics) while remaining homogeneously
colored and hence less conspicuous to another species
(say, its predators). How important this factor really is
remains to be determined, but a large number of coral-reef
fishes are either longitudinally striped or vertically
barred *(e.g.*, Chaplin, 1972).

Some of the possible solutions to the opposing needs
of concealment and conspicuousness are summarized in *table
7-V*. It is surely an incomplete list, but demonstrates
how consideration of optical principles in ecologically
relevant situations may explain certain characteristics
of animal signals.

A *Final Note on Determining Conspicuousness*

This chapter considered optical principles that promote visual conspicuousness. Each principle was then ta-

Table 7-V

Concealment and Conspicuousness:
Possible Solutions to Opposing Needs

Class of Solution	Specific Manifestations
timing	change color for communication display hidden coloration movement *vs* stationarity (see text)
background	special environmental display-sites different viewing point for receiver and other observer

addressee-specificity	specular reflectance to receiver via
	    body posture
	dichromatic reflectance to receiver via
	    piloerection
	iridescent reflectance to receiver via
	    sunlight-orientation

general orientation	display specific body part to receiver

distance	element-matching concealment from afar
	spatial fusion concealment from afar
	spatial-frequency fusion at non-optimum
	    distances and species-differences in
	    optima

ken to be an evolutionary hypothesis and, where possible, expectations derived from the hypothesis were checked preliminarily using an incomplete and informal comparative method *(table 1-III)*. In terms of the four determinants (causes and origins) of behavior articulated in *ch 1*, I have studied control factors and used them to generate hypotheses about preservation of signaling behavior (specifically, the adaptiveness of characters under the control of natural selection). Any ultimate understanding of conspicuousness is more complicated than this, however, because ultimate understanding must interrelate all four

of the biological determinants: control, ontogeny, preservation and phylogeny (*table 1-II*, p. 13). In this regard,
three final points may be noted.

First, individual animals may find by trial-and-error
those characteristics of their signals that promote detection by intended receivers. In ontogenetic terms, this
chapter was written as if differences in behavior and morphology of signaling were determined entirely by differences in the genes of the senders. In fact, signaling
differences could also be determined by differences in individual experience. In many cases, this seems unlikely,
but it is possible for an animal to try various gestures
and postures until one is noticed by its receiver, or for
an animal to assume a variety of colors until learning
which one attracts the most consistent attention in some
given situation. Only specific ontogenetic study can reveal how signaling behavior actually develops in the individual.

Second, species-specificity of signals in and of itself does not unequivocally prove that the signals are
being preserved in the population by means of natural selection. If specific learning or more general experien-
tail factors play a role in the ontogeny of signaling,
then the signal characteristics developed may show similarities from generation to generation for either of two
reasons: in each generation animals learn the same things,
or there is cultural transmission of signaling characteristics from older to younger animals in the population.
Therefore, generational similarities may be due to natural
selection of genetically transmitted traits, cultural selection of traditional traits or repeated individual learning from generation to generation.

Finally, animals may not show the optimum characteristics for signal-conspicuousness because of historical
constraints that are very difficult to study effectively.
In terms of genetically transmitted characteristics e-
volved through natural selection, historical constraints
may lie in the ancestral origins of signals. For example,
animals may lack the metabolic pathways for synthesizing
certain biochromes that would provide ideal coloration.
In terms of traditionally transmitted characteristics,
the origins of some signal-characteristics may have occurred
in the distant past; the characteristics continue to be
taught new generations, not because they work best, but
because they work adequately enough to inhibit experimentation with new ways of signaling.

Overview

The signal-to-noise ratio in the optical channel depends primarily upon how visually conspicuous a signal
appears against its background and how well it can be detected with a less than transparent medium between sender
and receiver. Partly by "reversing" principles of deception articulated in *ch 6*, it is possible to identify some
general principles of conspicuousness, and then to search
for ways in which these principles are exhibited by signals in different environmental situations. The major
variables affecting conspicuousness are the light available for reflection by the sender, the optical nature of
the background against which the sender is viewed, and the
types of optical disruption signals suffer during transmission. In each case, it has been possible to identify
at least some of the environmental situations that differ
consistently in these variables, and to hypothesize what
characteristics optimum signals should have as a result.
There are, however, many unsolved problems, even in this
preliminary survey. Finally, certain optical strategies
have been uncovered whereby a sender can appear conspicuous to its intended receiver, yet remain concealed from
undesirable observers such as predators. Conspicuousness
is achieved by both behavioral and morphological elements
of signals, but in general the latter seem more important
than the former.

Recommended Reading and Reference

As far as I have been able to determine, no one has
written a major work specifically aimed at articulating
how animals maximize conspicuousness under different ecological conditions. References on concealment in *ch 6*
might be consulted with this viewpoint in mind.

Chapter 8

INFORMATION

*The literature of semiotics is thus replete
with mere restatements rather than solutions
of problems, and the need for different kinds
of theory at different levels of "coding" appears a most pressing task.* --Sebeok (1968)

Sebeok (*e.g.*, 1965, 1968) partitioned semiotics--the
general study of signs--into *zoosemiotics* and anthropo-
semiotics, and further separated the latter into linguistic and non-linguistic divisions. The distinctions are
useful for this chapter, which attempts to relate the
characteristics of non-linguistic optical signals to the
information they transfer. Semiotics owes to Charles W.
Morris (*e.g.*, 1946) the insightful division of problems
into semantics, syntactics and pragmatics, one of the many
triads of the semiotic literature. This triad has been
used frequently but in varying ways by subsequent authors
(*e.g.*, Cherry, 1957; Weaver in Shannon and Weaver, 1949;
Marler, 1961; W.J. Smith, 1968), as mentioned in *ch 2* (pp.
48-49). Translating the triad into the practical language
of physical signals, *semantics* concerns the relations of
signals to their referents, *syntactics* the relations of
signals with one another, and *pragmatics* the relations of
signals to their consequences. The objective of this
chapter is to see whether exploration of these aspects of
informational transfer help define the characteristics of
optical signals.

Semantics

Signals encode information by specifying a subset

-249-

___

from the set of alternatives about which the receiver is
uncertain (*ch 2*). The subset specified is the referent
of the signal, but the referent need not be a physical
object: it may be something quite abstract, such as a vector of probabilities of future behavioral acts of the sender. Therefore, in considering the semantic problem of
how signals relate to their referents, one is constrained
to study cases in which the referents are known. In most
cases, the referents are particular animals, objects,
places or their observable attributes, such as sex of the
animal, type of object, or potential use of a place (*e.g.*,
for a nest).

The American philosopher Charles Sanders Peirce,
founder of semiotics, recognized about 1906 that signs
may be classified as indexes, icons and symbols. This triad serves a useful organizations device, and as the names
suggest, *indexic* signals point out their referents, *iconic*
signals look like their referents, and *symbolic* signals
stand for their referents in any way held in common by
sender and receiver. The semiotic triad refers to properties of signals, since a given signal may exhibit more
than one of these properties.

indexic properties

An index carries information by literally pointing
out its referent, so one may ask how optical signals might
be constructed to act as spatial pointers. In terms of
signal-elements (*table 4-III*) the question becomes whether
orientations, movements, shapes, structures and reflectance patterns may be used to index referents.

The mere body *orientation* of a sender is widely used
by animals as an indexic signal pointing out various referents. The commonest referent is the social companion
this is being addressed (the intended receiver). The orientation adopted by the sender to point out the addressee
may vary according to the anatomy of the communicants, to
the other information being transferred at the same time
and to many other factors (*e.g.*, *table 7-V*). In fact,
pointing out the addressee is also a syntactic problem,
to be considered below.

Orientation with respect to referents other than the
addressee is also common. For example, an animal may give
some particular signal when sighting a potential predator,
and then visually track the predator sighted. Social com-

panions may then look at the sender's direction of gaze in order to locate the predator, so that the direction is indexic (*e.g.*, Hall and DeVore, 1965; Struhsaker, 1967). One may expect accompanying morphological structures or color patterns to be arranged so as to make this indexic signal more conspicuous and unambiguous. It seems possible that the head-striping of many small songbirds serves such accentuation of the indexic direction of gaze, since regardless of a viewer's position relative to the bird, the directions its head is turned becomes quite obvious *(fig 8-1a)*.

Fig 8-1. Coloration to enhance indexic orientations. The convergent longitudinal markings on the head of the lark sparrow (a) facilitate recognition of its direction of gaze, and the markings on the pinnae of jackrabbits (b) facilitate detecting their orientation with respect to a sound-source.

It is largely irrelevant whether the sender fixes the object of gaze monocularly or binocularly, so long as the

receiver knows which and can reconstruct the probable direction of gaze from seeing the position of the sender's
head. If stripes were to run from one side of the head
over the cap to the other, instead of running longitudinally to converge at the bill, the indexic information
would be more difficult to recover; such a pattern, which
would be disruptive (ch 6), is rare among birds. Other
potential functions of head-striping in birds were discussed in ch 4 in relation to hiding the eye, preventing
glare, providing sighting lines, etc.

     Years ago, Haartman (1957) drew attention to indexic
signals of hole-nesting birds, which display at the nest-
site. He noted that birds such as woodpeckers and nuthatches tend to have dorsal signal-markings, since the
backs of these birds are visible to mates when they are
pointing out possible nesting holes.

     S. Witkin (pers. comm.) notes that in hole-nesting
and in other birds presumed indexic coloration is frequently triangular in shape. For example, hole-nesting
chickadees (Parus) have a prominent white triangle on the
cheek, with the apex pointing toward the bill. The head,
as seen from above, displays a similar black triangle,
and the chin as seen from below the bird also displays a
black triangle. The dark head-patterns of many other birds
when seen from above also form a triangle pointing toward
the bill. Therefore, triangularity (as well as the kind
of striping illustrated in fig 8-1a) may be an important
indexic color pattern in animals.

     Orientation of an appendage, as opposed to gross or-
ientation of the head or body, sometimes constitutes an
indexic signal. The commonest example is the orientation
of the pinnae in various mammals: the pinnae are oriented
toward sounds, particularly unusual sounds that might con-
note danger, and so a companion viewing the listening animal may be able to approximate the direction of the sound
even if he himself did not hear it. It is common to find
special coloration on mammalian pinnae (e.g., black tips
on the ears of lagomorphs, fig 8-1b), although like head-
striping in birds these marks may be involved in other
communication as well. In some cases, special structures,
such as the eartufts on some felids, also accentuate the
indexical quality of pinnae-orientation (see figure 8 in
Hailman, 1977a).

     Movement, by definition, involves a change in position
so that the directional components of movements could

serve as indexic signals. For example, the response of
some social animals is to flee a potential danger, so that
the direction of fleeing indexes the direction of danger
as being immediately behind. In such cases one expects
posterior, rather than anterior, enhancement by color pattern, and this is commonly the case (*e.g.*, the white-tailed
deer's white tail is raised when fleeing: see figure 20
in Hailman, 1977a).

Movement of part of an animal, as opposed to total
body displacement in space, may also be indexic. Statements by Lorenz (*e.g.*, 1941) to the effect that the inciting movement of female mallards is rigidly fixed and un-
oriented notwithstanding, quantitative study shows clearly
that inciting is an indexic signal (Stillwell and Hailman,
in prep.). The female turns her head toward an intruding
male, stopping the movement when looking binocularly directly at him, then returning her head to straightforward
and repeating the directed inciting motion (*fig 8-2a*). It
is instructive to note that, except for usually hidden
wing-markings, the female mallard is nearly uniformly a
concealing brown color. The one patch of signal-coloration visible while swimming is the orange-and-black marking of her bill--the pointer used in the inciting signal.

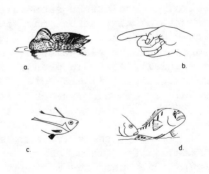

*Fig 8-2. Indexic signals
include movements, such
as the female mallard's
inciting (a), in which
her orange bill is moved
to point toward an in-
truding male; shapes, as
in pointing with the in-
dex finger (b); struc-
tures, as in the gill
covers of certain male
fishes (c) that index
the position of the fe-
male for spawning; and color-patterns, such as the markings on the anal fins of male mouth-brooding cichlids (d)
that direct the female to the proper place to take the
male's sperm into her mouth.*

The *shape* of an animal is a behavioral signal-element
created by adjustment of body parts relative to one an-

other *(ch 4)*. Although orientation of the head *(e.g.,* *fig 8-1a)* is strictly speaking a kind of behaviorally created shape, ethologists usually separate such cases from more stereotyped postures that create less common shapes. For example, the open hand is not immediately taken by most persons as an optical signal, unless its orientation toward someone indicates it is indexic greeting. However, a fist with the index finger protruding is a special shape that is immediately taken to be an indexic signal *(fig 8-2b)*; the receiver confirms or refutes this communicative hypothesis by looking in the direction toward which the finger is pointing to see if something special is there. Indeed, pointing with the index finger is so familiar that it might be called our archetype optical index, yet such signals seem surprisingly rare in animals. Chimpanzees also use a finger indexically, although perhaps only after contact with humans. Hunting dogs appropriately called pointers index prey by a special body posture in addition to the orientational component of their direction of gaze.

In sum, behavioral elements used as indexic signals are often accompanied by patterns of coloration that enhance the indexic quality. Examples are summarized in *table 8-I*.

Special *structures* evolved for indexic functions in animal communication seem rare, although as noted above structural adaptations may enhance the indexic qualities of orientations, movements and postures. Nelson (1964) notes that the glandulocaudine fish *Corynopoma* has evolved a special "courtship paddle" that is extended for the female to nip at *(fig 8-2c)*. The position of the end of the paddle in a sense indexes the proper spatial position of the female for spawning relative to the male, and hence is a special indexic structure.

Finally, there are many examples of indexic *color-patterns* of animals, apart from coloration to enhance orientation, shape and movement elements of behavior. For example, in some mouth-brooding cichlid fishes the young fry return to the dark gape of the parent and when older are attracted to stay near the closed-mouthed parent by attempting to enter a dark spot on its side. The spot is thus initially a deceptive mechanism inducing the young to stay close while they presumably learn optical, chemical and other characteristics of the parent. Another example from cichlid fishes is the marking on the male's

Table 8-I

Behavioral Indexes

Indexic Element	Specific Examples	Enhancing Characteristics
orienta-tion	direction of gaze	convergent striping triangular pattern
	pinnal orientation	coloration or hair-tufts on tip
movement	fleeing direction	posterior marking
	move toward referent	coloration of pointer
shape	point to referent	convergent shape

anal fin in *Haplochromis burtoni* (Wickler, 1968). After laying eggs, the female takes them into her mouth, and the male lies near the substrate while emitting sperm. The female puts her mouth near the indexic fin-spots and ingests the sperm for fertilization of the eggs in her mouth cavity *(fig 8-2d, p. 253)*. Wickler believes the spots to be egg-mimics that deceive the female, who attempts to pick up the "eggs" and thereby picks up sperm; however, this interpretation is not the only one possible (see *ch 9*).

In sum, indexic signals are widespread in animals, being found in both behavioral and morphological elements of optical signals of all kinds. Although not mentioned specifically here, extrinsic signals also may be indexic, and examples are provided later in this chapter. In some cases, knowledge of how indexic signals work leads to expectations concerning color patterns that enhance the indexic qualities *(table 8-I)*, and it would seem fruitful to give such patterns more analytical attention in future studies.

iconic properties

An iconic signal carries information by physically

resembling its referent, so one may ask how optical signals might be constructed to resemble other things. Although they were not identified by name as iconic, many
of the deceptive signals cited in *ch 6* are iconic in nature: all mimicry, for example, is iconic at least in a
derivative sense. For example, if a bird pecks at a moth,
which then spreads its underwings to reveal two large,
dark spots, then the moth is said to be mimicking the eyes
of some large animal that would frighten the bird. The
spots are an icon for eyes because they look to us like
eyes, although the bird presumably *mistakes* them for eyes.
Therefore, it becomes useful to separate deceptive mimicry
(signals actually mistaken for other things) from general
mimicry or iconic signals in the restricted sense (signals
recognized as signals, which look like something else).

Iconic *behavioral signals* (especially movements and
postures) are almost universal in animal optical communication, although one might not readily recognize this fact.
When a gull, for example, adopts the upright posture (the
"threat" illustrated in figure 24 of Hailman, 1977a) the
signal looks like the start of a peck at the opponent and
has probably evolved from just such a behavioral pattern.
When the gull pecks into the grass, the movement looks
like fighting, even though it does not involve an opponent;
ethologists call this "redirected aggression." Such redirected movement is an icon to the opponent on the adjoin-
ing territory, another kind of "threat" signal. In these
cases of iconic signals, the referent is the behavior that
the sender might engage in subsequently, say if the opponent were to move closer to the sender. Marler (*e.g.*, 1977)
has called such icons "predictors." It is generally be-
lieved among ethologists that the overwhelming majority
of behavioral signals are evolved from intention movements
of non-communicative behavior (see references in Hailman,
1977a and *table 4-IV*, p. 107). Therefore, all such signales are probably iconic to some extent--they look like
other behavior that the sender is broadcasting as probable
in the immediate future.

It is useful to point out that iconic behavioral signals do not all have to evolve through the process of phylogenetic ritualization. The form of iconic signals may
also develop through experience of the individual sender
(*table 1-II*). A boy may learn, for example, that clench-
ing his fist is noticed by a playmate as preparation for
hitting; after that, the boy may purposely close his fist

in order to communicate, consciously or subconsciously,
his willingness to fight. In both phylogenetic and ontogenetic developments of iconic signals, the receiver already cues upon preparatory or "intention" movements of
some act; then natural selection or individual experience
can proceed to enhance the signal-value of the icon to
make it more conspicuous and unambiguous.

The question is therefore not whether individual movements and postures can be icons, since many of them clearly
are. Rather, the question is whether orientations, morphological specializations and other signals such as whole
complexes of behavioral signals are also iconic. In the
case of *orientation*, it is difficult to find an example of
a pure icon. Orientation, say toward the addressee, may
accompany an iconic movement or posture, but in this case
it is only indexic--unless one cares to argue that the
orientation of the icon is itself iconic for the orientation of the icon's referent. For example, a threat posture may be an iconic signal for possible attack and the
orientation of the posture toward the opponent indexes the
intended receiver of the signal. Since the receiver is
also the intended recipient of attack, the orientation of
the signal itself may be taken as an icon for the directed-
ness of the attack.

Structures and *color-patterns* are frequently iconic.
Many mammals have special warts or other structures that
look like fighting weapons such as teeth or horns (see
figure 7 in Hailman, 1977a), and others have color-patterns
on the ears or elsewhere that also look like weapons (see
figure 22 in Hailman, 1977a). These structures and patterns have been called "automimetic," although it is doubtful that they are truly deceptive in the sense of *ch 6*.
Opponents probably do not perceive these iconic signals
as actual additional weapons. However, because opponents
do recognize the display of weaponry itself as signaling
probable fighting, the repetitious nature *(table 7-I)* of
the added weapon-icons enhances the signal by increasing
redundancy.

There is an interesting reputed case of "automimicry"
in which coloration creates the iconic resemblance, but
the color-pattern itself is unexplained. The male mandrill has a red nose and blue cheek skin that resembles
the red penis and blue scrotal area. Wickler (1968) cites
primate examples in which anogenital display is used in
social, rank-related signaling, including cases in which

males develop female-like swellings that are used as social signals. Presumably the mandrill's face mimics genital signals, in effect repeating them anteriorly for use
in face-to-face signaling, but the coloration of neither
genital nor facial areas has been well explained.
      Jouventin (1975) calls the coloration "luminous," and
points out that it is conspicuous in the thick equatorial
forest. This possibility is supported by the fact that
red and blue are both complementary to green colorations
(*fig 7-3*, p. 203), and hence might contrast well against
various backgrounds of green vegetation. Another possibility is that the coloration makes use of the depth-illusion based on color (*fig 5-18*, p. 152), and hence enhances the basal relief of the signal-morphology. Because
red is seen as closer than blue or violet, the penis may
appear to stand out from the blue scrotal background. A
blue object would not protrude from a red background, so
the hypothesis provides a clear prediction concerning
coloration to enhance relief of morphological signals.
      Extrinsic signals are natural candidates for iconic
representations, but they seem rare in most animals. Every
drawing of an animal in this book is an icon, but animals
themselves rarely draw pictures. Many iconic-like signals
of animals appear to involve true deception. For example,
male of some species of empiid flies present prey to the
female before copulation, which occupies her sufficiently
that she may not eat the male. In other species, the male
first wraps the prey in silk, and in still others presents
an empty silk balloon (Kessel, 1955). In the last case,
the female may be deceived and actually search for prey,
but it seems more likely that the balloon has become symbolic of the male's copulatory intentions; by either interpretation, though, the balloon is not really an iconic
signal. In deceptive mimicry there are examples such as
the Asiatic spider that builds spider-like masses in its
webs (*fig 6-4*). Like moth eye-spots, however, the receiver
actually mistakes the stimulus for something else rather
than perceiving it as a signal that stands for something
else.
      Waxing speculative, a fascinating bit of natural his-
tory might hold an example of an extrinsic icon. The garibaldi is a Pacific fish growing to about 30 cm in which
the male is bright orange. He holds a year-round territory on to which he attracts females during the breeding
season, partially by culturing a 15- to 40-cm oblong patch

of red algae of various species. The patch is thus about
the same size as the fish, and the patch acts as a nest-
site, with eggs being attached to the algae by short filaments (T.A. Clarke, 1970). Is the male's coloration e-
volved as an iconic signal relating to the nest-site, or
is the nest-site an icon of the male? Or is the similarity
of shape, size and color purely evolutionary coincidence?

symbolic and mixed properties

A symbolic signal carries information by "standing
for" its referent in any manner held in common by sender
and receiver. For present purposes it may be taken as any
relation between signal and referent that is neither indexic nor iconic. Sometimes symbols are called arbitrary,
but I have avoided that word as being ambiguous.

"Arbitrary" means established by caprice rather than
reason, and in most cases one does not know how symbols
became established in animals. The actual relation between a symbol and its referent may be arbitrary in the
sense of having no specifiable connection (such as the signal pointing out or resembling the referent), but there
may be good historical reasons for having established the
relation originally. For example, the symbol "π" for the
ratio of the circumference of a circle to its diameter is
"arbitrary" in the sense that it bears no obvious relation
to the words quoted as its referent, to a drawn circle, or
even to the numerical value of the ratio itself (3.14159+).
However, the relationship was not *established* arbitrarily.
The symbol is the 16th letter of the Greek alphabet, the
initial letter of *peripherion* (periphery), and hence is
very nearly a simple abbreviation. Analogously, many of
the various symbolic elements of behavioral signals of animals probably evolved from behavior precursors that are
obscure today. The signals are neither indexes nor icons,
hence they are symbols, but their historical establishment
was not necessarily arbitrary in any useful sense of the
word.

It is evident that many behavioral elements of signals (orientations, postures and movements) are symbolic,
as are many morphological elements (structures and color-
patterns), since the only operational criterion is that
they be neither indexic nor iconic. For example, the adult
male common flicker has a mustache mark lacking in the fe-

Table 8-II

Semantic Properties of Optical Signals

Signal-elements (re table 4-III)	Semantic Classification indexic	iconic	symbolic
orientations	XX	0?	X
movements (incl. gestures)	X	XX	X
shapes (incl. postures)	X	XX	X
structures	X	X	X
reflectance patterns	X	X	X
extrinsic signals	X	X?	X

XX - *particularly common*, X - *occurs*, 0 - *rare or absent*.

male (figure 10 in Hailman, 1977a), so that the presence
or absence of this signal symbolizes the sex of the bird.
Whenever the referent is something that cannot be pointed
our in a literal sense or represented iconically, the signal that stands for it *must* be a symbol. Whether there
are useful classes of symbols to be distinguished remains
an open question. The present point is simply that by
discovering that a signal is symbolic tells one nothing
about its characteristics.

A summary point to be made about the semantics of optical signals from the viewpoint of their characteristics
is that a given signal or complex of signals can exhibit
indexic, iconic and symbolic properties simultaneously.
An example suffices to make the point. The male stickleback builds a tunnel-like nest and courts a female with
various signals, one of which is "creeping through" the
nest while the female watches. The action is indexic because it points out the nest in space and iconic because
it resembles what the male's actions will be if the female
deposits eggs in the nest: he will swim through the nest,
depositing sperm as he goes. Other stickleback courtship
actions, and perhaps aspects of creeping through itself,
are purely symbolic, having no features that point out
something or resemble something else.

Table 8-II summarizes the kinds of signals that are known to be indexic, iconic and symbolic. Finally, it is useful to recall that these sections on semantics concern only the classification of relations between signals and their referents, and the implications these have for the characteristics of optical signals. I have not been concerned with semantics in the dynamic sense of the communicative act, which is a broader concern and tangential to the specific aim of this volume.

Syntactics

Signals carry information by designating referents, but the referents may be designated by whole complexes of signals, as well as individual, isolated signals. Therefore, the kinds of signals grouped together, the order in which they are exhibited and the temporal and spatial relations of the signals within a complex all may affect the information transferred. Syntactics is the study of how signals relate to one another in transferring information, and the syntactics of animal communication may turn out to be one of the most difficult and challenging areas for study.

As in the previous major section on semantics, however, I am concerned here only with how exploration of the problem-area may help to clarify the observable characteristics of optical signals. I find at least three classes of syntactic problems: coding, destination and interpretation. These are defined and discussed in turn.

coding problems

Certain relationships among signals may be governed by considerations of the communicative process itself, which I shall call syntactic problems of *coding*. This area has received little attention, and I consider just one example each of problems relating primarily to the sender and to the receiver.

Attneave (1954) and later H.B. Barlow (1961a,b) drew attention to the coding principle Barlow calls "message compression," in which it is reasoned that common sensory inputs should be represented neurophysiologically by brief neural events requiring little energy. Dawkins (1976) has extended this notion to motor outputs of animals, and it

seems reasonable to extend it further to signal-production.
Under this principle of *signal-economy*, commonly used signals should be brief and require little energy for production whereas longer and more costly signals should be
rarer.   Zipf (1935) pointed out that word-length was inversely related to frequency of occurrence, but I cannot
find a quantitative example from animal optical signals.
[This principle is not to be confused with Zipf's "law"
or "principle of least effort," which states that log
frequency of occurrence is inversely proportional to fre-
quency-rank.   Mandelbrot (1953) showed that relation to be
a mathematical consequence of dividing time randomly into
segments.   Moles (1968) suggested that departures from
this random expectation will prove to be the instructive
aspects of analyzing communication, and Schleidt (1973:
367) has provided an example.]   It is a common impression
among ethologists, though, that frequent signals of animals
are those of short duration, so the notion of signal-economy provides a useful hypothesis for future studies.

From the receiver's viewpoint there may be limits on
the total number of optical signals that can be reliably
discriminated, and such limits should affect the signals
used by senders.   At least two factors are relevant to
signal-discriminability: the total number of different
signals and how easily each is discriminated from the others;
Moynihan (1970) and Johnston (1976) explore this topic in
some detail.   One may, for instance, be able to discriminate a hundred different geometrical shapes, given juxta-
position and adequate viewing time.   However, if these
shapes are seen in succession rather than simultaneously,
and for short durations, confusion may ensue.

There is no limit to the amount of information that
can be transferred by the presence or absence of one sig-
nal-type; for example, all real numbers can be represented
in binary notation (*1* and a place-holder for its absence,
for which *0* is used).   The disadvantage of the binary system is that a long string of numerals may be required to
denote a number that can be written more parsimoniously
in base-ten (*e.g.*, 111111 in binary is simply 63 in deci-
mal notation).   Numbers could be written more parsimoniously yet with 20 or 50 different numerals, but such a large
set might prove unwieldy because so many visual discriminations would be required.

Smith (1969) pointed out that signal-repertoires were
often limited in size.   Moynihan (1970) attempted to com-

pile the signal-repertoires of various vertebrate species,
finding fishes to have 10 to 26, birds 15 to 28 and mammals 16 to 37 displays, depending upon the species. Wilson (1972) reports that analyses by C.G. Butler and himself show social insects to have between 10 and 20 displays.
These figures are based on totals for all modalities, not
merely optical signals, and they are restricted to ritualized signals *(ch 2)*. What constitutes a ritualized signal is difficult to judge objectively, and in any case
these numbers probably underestimate the actual repertoires
of information-transmitting behavioral patterns. For example, Moynihan's (1970) compilation of Altmann's data for
the rhesus macaque yields 37 displays whereas Schleidt (1973)
recognizes a repertoire of 59 "elemental key sign types"
based on the same series of studies. When the behavioral
units overlap in time they may be recognized by receivers
as a new type, and of course the role of environmental context *(ch 2)* cannot be ignored. The point is that objectively defining the repertoire is difficult if not impossible. However, by using the same criteria it may be possible to compare repertoires of related species to see
what factors dictate the size of signal repertoires.

One way to decrease signal-ambiguity, and hence allow
for more types of signals and more parsimonious coding,
is by low variability in the signals themselves. Color-
patterns and morphological structures may show low variability from individual to individual, and from time to
time in the same individual. Behavioral signals, on the
other hand, and color-patterns that are very labile, may
vary from performance to performance, as well as among
different individuals of the same species, and so are especially prone to variability as signals. Ethologists have
long recognized that signal-movements are more *stereotyped*
than other behavior, but because of the technical difficulties of measuring stereotypy by means of motion picture
film, often obtained under trying field conditions, quantitative studies have been slow to appear. I recall the
bitter cold winds endured by Dane and Walcott during their
pioneering studies of goldeneye displays filmed off the
Massachusetts coast in winter (Dane *et al.*, 1959; Dane and
van der Kloot, 1964). Film studies of other avian signals
(Hailman, 1967a; Wiley, 1973) and signals of other animals
such as lizards (*e.g.*, Jenssen and Hover, 1976) are now
available, but it is still too early to make meaningful comparisons among species (see G.W. Barlow, 1977). Stereo-

typy may also be important in conspicuousness *(ch 7)*.

In future studies of stereotypy of optical signals it will become increasingly important to distinguish the information-carrying sign-vehicle *(ch 2)* from the inclusive signal. Golani (1976) has shown that in what appears to be highly variable behavior in social interactions in mammals, certain aspects of the behavior are not at all variable. Although his examples deal with behavior that involves primarily chemical and tectile stimuli, the same principle may apply to optical signals, and needs to be given experimental scrutiny.

It follows that if signals are to have low variability to prevent ambiguity, they should also be as different from one another as possible. The extreme case was recognized by Darwin (1872), who pointed out that signals arising from "opposite emotions" should have "opposite" characteristics. Oppositeness is a difficult quality to define objectively, but the general notion of Darwinian *antithesis* is widely documented in animal signals, particularly in threat and appeasement signals (see figures 23 and 24 in Hailman, 1977a). Baylis (1974b) illustrates colorational antithesis in the fish *Neetroplus nematopus,* in which the social, schooling fish is gray with a black bar and the solitary, breeding territorial fish is gray with a white bar.

No discussion of economy, repertoire, variability and discriminability of signals is complete without a reminder that the environmental context of communication *(ch 2)* is a potentially important factor in any signaling. For example, repertoire size need not be so great when differences in context are the primary information-carrying aspects of communication, and the important aspects of discriminability may be among contexts rather than among signals.

destinational problems

Certain signals or aspects of signals may determine the intended destination or addressee of other signals. Such signals are therefore semantically indexic, but also raise syntactic problems because their information relates to other signals for which they point out the addressee.

Simple *approach* to within signaling distance or transmission at some particular *time* may help determine an addressee. Here I am more concerned with particular aspects

of signals, especially *orientation* with respect to the intended receiver. By being able to see the addressee, the
sender can insure that his signals are sent in that direction and that the optical channel between him and the destination of his signals is clear. If the animal is sufficiently large, he may not be able to tell whether the
addressee can or cannot see some part of his body distant
from his own eyes. Therefore, one may expect *encephalization* of optical signals to be favored under these conditions: large animals living in environments with opaque
noise *(ch 7)* when specific addressees are intended. It is
therefore not surprising that trunk postures of elephants
and facial expressions of equines are important optical
signals in social interactions (Tembrock, 1968). A different factor favoring encephalization was mentioned previously in a different context. Because sense organs are
directed toward objects of interest (*e.g.*, potential predators), and sense organs themselves are encephalized, one
expects to find encephalization of indexic signals in general.

A topic that follows directly from addressing the signal is monitoring its reception by the addressee. If the
sender can see the addressee's eyes, then the sender knows
that at least his eyes can be seen. *Eye-contact* is therefore a reciprocal signal in and of itself, one whose syntactical aspect relates to the signals of the other communicant. E. Burtt *(pers. comm.)* reminds me that eyerings
of birds are important signal-markings for locating the
eyes. Because the iris color of most species is very dark,
birds with light head coloration have no special eyering
coloration, whereas species with dark heads have contrastingly light eyerings that serve to locate the eyes.

In turn, the act of breaking off eye-contact by rotating the eyes or the entire head or body so as to look
elsewhere becomes a dramatic signal indicating non-recep-
tion. These have been called *cut-off signals* (Chance, 1962),
but they are not in all cases necessarily due to syntactic
considerations. For example, "facing-away" or "head-flagging" of gulls, in which the members of a mated pair simultaneously turn their faces away from one another, also
results in their hiding weapons, namely bills (Tinbergen,
1959), and hence may carry other meaning than breaking eye-
contact: they demonstrate antithesis of threat.

Extrinsic signals may also be used to index the addressee. In many species of puddle-ducks the male per-

forms a "grunt-whistle" display, in which his body rises
out of the water, but his neck is arched downward with
his bill in the water (e.g., Lorenz, 1941). The lateral
movement of the bill sends an arc of water flying, and
McKinney (1975) showed that males perform the grunt-whistle
only when spatially aligned such that the water-jet moves
directly toward the female. Some of the preliminary shak-
ing movements that precede the grunt-whistle may be involved in positioning the male and aiming in preparation
for displays (see Simmons and Weidmann, 1973).

interpretational problems

Some signals provide aids to the interpretation of
others by the receiver. Again, little attention has been
given this subject in relation to optical signals, but at
least three kinds of interpretation have been reported:
behavioral context, punctuation and quantification.

A signal concerning *behavioral context* tells the receiver how to interpret, in a general sense, an entire
group of signals. The best-documented and most discussed
example of contextual syntax in optical signaling is the
case of special "play signals" which G. Bateson (1956)
called "metacommunication." Metacommunication is, strictly
speaking, communication about communication, so that any
signal with syntactic content really qualifies as metacommunication. It remains to be seen whether the broad or
restrictive usage will prevail, and perhaps the term is
best dropped as already ambiguous. At any rate, Bekoff
(1975) has shown that coyotes may flex the forelegs, extend the hindlegs and wag the tail *(fig 8-3a)* before giving other optical signals such as baring the teeth, flattening the ears and erecting the mane. When the former
"play" signals precede the latter "threat" signals, the
receiver shows a low probability of submissive behavior,
but when "threat" is given without the contextual syntax,
the receiver shows a high probability of submissive behavior.

Contextual signals may also be used deceptively. In
fact, such feigning is probably the chief category of intraspecific deception. For example, Carpenter (1964) notes
that monkeys use sham feeding in order to get close to a
conspecific for performance of behavior unrelated to feeding: as an introduction to play (p. 27) and by males as an

approach to females for copulation (p. 61).

There are several nomenclatural and definitional problems concerning interpretational signals that are worth considering briefly. First, behavioral context discussed above refers to the general kind of activity being performed by the sender; environmental context discussed in *ch 2* refers to information coming from sources other than the sender that helps decode the sender's signals. The latter is a more general concept, which may provide information about the sender's behavioral context or about other aspects of his signals. Second, general terms such as metacommunication (Bateson, 1956) or grammar (Schleidt, 1973) seem best used as synonyms for syntax, since it is useful to separate various kinds of syntactic signals. Lastly, there may be no firm line between signals denoting behavioral context, discussed above, and those that direct how other signals shall be read in a more structural sense.

Punctuational signals are those that specify how other signals should be read in a structural sense. For example, although tail-wagging in mallards *(fig 8-3)* has several

Fig 8-3. Syntactic interpretational signals. The posture of the coyote (left) denotes the behavioral context of "play" and the tail-wagging of the mallard (right) may be a punctuational signal announcing other displays to follow.

non-communicative functions, it occurs with high probability just before and after sequences of pure display behavior (Hailman and Dzelzkalns, 1974). Its sequential placement therefore can transmit syntactic information similar to that of a capital letter at the beginning of a sentence and a period at the end: punctuation. Jenssen and Hover (1976) report similar small head movements of anoles given before, after or between qualitatively different segments of primary display. Such movements prior to display may also be indexic in that they attract visual attention of receivers. Schleidt (1973) discusses walking toward and walking away from receivers in rhesus macaques as signals that initiate and terminate behavioral interaction. All these punctuational signals share a characteristic identified by Bekoff (1976) in signals of play context: they are relatively frequent and of relatively short duration. The correlation between these two variables is expected from the principle of signal economy, discussed previously.

The final class of interpretative signals in optical communication is quantifying aspects or "signal intensity. It seems possible that *quantification* is merely a subset of a broader set of signals and aspects of signals that modify others, analogous to the way in which adjectives and adverbs modify nouns and verbs. However, quantifying aspects seem to be the only clear modifiers reported in optical signaling. The most detailed study of signal intensity is Brown's (1964) work on the Steller's jay, which raises its crest to various angles (see figure 12 in Hailman, 1977a). Brown documented quantitatively the degree of crest-raising in various social situations and found it to correlate primarily with "increasing resistance of the opponent" in agonistic encounters. This finding suggests that the degree of crest-raising quantifies something like the probability of attack by the displaying bird.

There has been much discussion among ethologists as to why quantifying signals show continuous or discrete variation, ever since D. Morris (1957) pointed out that some signals showed the latter pattern, which he called "typical intensities." Sometimes continuous variation is referred to as analog, and discrete variation as digital (*e.g.*, G.W. Barlow and Green, 1969). J.L. Brown (1975) has criticized such usage and prefers graded and discrete. In fact, discrete variables of mathematics may be graded (show ordinal or higher level variation), and the most straightforward language seems to be simply that of mathematics: continuous and discrete (able to take on any value or able to take on only certain values along a continuum). D. Morris (1957) suggested that discrete variation or "typical intensity" was useful in standardizing a few points on a continuum that the receiver could readily distinguish (see discussion above under *coding*). However, no discrete system of variation has been investigated with the rigor of J.L. Brown's (1964) study of continuous variation in crest-raising, and it may turn out that discrete systems involve factors not heretofore considered.

It seems worth emphasizing a point that is obvious
upon thought, but might escape notice: not all signals can
be readily quantified. For example, the male cardinal is
red and has no physiological mechanisms for creating degrees or shades of redness in behavioral interactions.
Fishes, with their melanophoric systems that respond rapidly *(ch 4)*, can show degrees of coloration. Many shapes,
such as the crest-erection of the Steller's jay, can show
degrees, and most movements can be quantified by amplitude
or frequency, or both. When pragmatic needs of signaling
require quantification, certain signals may therefore be
evolved more readily than others. Furthermore, when two
or more qualitatively different kinds of signals are to
be superimposed, and each quantified in some way, there
are further constraints on the choice such that the full
ranges of quantification of both signals are compatible
(fig 8-4).

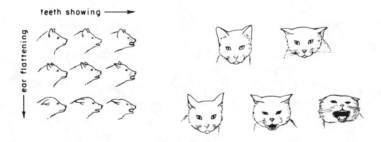

*Fig 8-4. Superimposability of signals. In canids (left)
opening of the mouth and flattening of the ears may be
varied independently to create all possible combinations
(after Tembrock, 1968, from Eibl-Eibesfeldt). In the domestic cat (right) several signal-elements may be superimposed, including flattening of the ears, rotation of
the ears, dilation of the pupils, opening of the mouth,
etc. (modified from Leyhausen, 1956).*

Some of the syntactical problems of optical signaling
are summarized in *table 8-III* (next page), which indicates
suggested solutions or expected signal-characteristics for
specific problems.

Table 8-III

Syntactic Considerations of Optical Signals

Syntactic Problem-areas	Specific Problems	Possible Solutions and Signal-characterestics
coding	transmission-efficiency	signal-economy
	signal-discriminability	stereotypy antithesis
destination	channel clarity	indexic address
	signal-visibility	encephalization
	receiver attention	eye-contact
interpretation	behavioral context	"metacommunication"
	punctuation	short duration
	quantification	signal "intensity"

Pragmatics

Pragmatics, the study of the consequences of signals, is the most central and perhaps most difficult aspect of animal communication. It is natural to focus on the effect communication has on the receiver and thus the benefits derived by the receiver who gains information from the communication. In terms of evolutionary theory, however, the central pragmatic problem concerns the benefits derived by the *sender*. In some cases, it is relatively easy to suggest how a sender benefits from signaling. For example, a threat signal may secure for the sender access to some resource without having to fight for it. Recently, theoretical attention has been given to advantages of deception (*e.g.*, Wallace, 1973; Otte, 1975).

In other cases, the benefits to the sender must be quite indirect. Current theory emphasizes the importance of correlated genotypes among social animals (J.L. Brown,

1975; Wilson, 1975). In simplified terms, an animal may promote the survival of his own kind of genes in the population by helping other animals who carry copies of his genes. The degree of help is a function of the cost to the first animal, the benefit to the second, and the degree of genetic relatedness between them. It is not feasible in this book to scrutinize this subject of evolutionary "kinship selection" and "altruistic behavior," except to note that much more needs to be learned. The major concern here is to see if the type of communication, as it is observed under natural conditions, helps to structure the characteristics of optical signals.

In order to investigate pragmatic aspects of signals, it is necessary first to provide a tentative scheme of types of relevant behavior. Then each of the major categories named is preliminarily surveyed to see if there are widespread characteristics that optical signals might have in order to bring about successful communication concerning the behavior discussed.

a behavioral classification

It proves convenient in devising a tentative pragmatic classification for communication to follow more or less closely the implicit functional classification used by ethologists to organize animal behavior conceptually. In terms of behavior in which communication occurs, there are three broad functional categories: agonistic, reproductive and cooperative behavior. These, in turn, may be subdivided for closer scrutiny.

Agonistic behavior is a peculiarly ethological term denoting behavior that involves a contest over something, such as food or access to a female. (Its root is the Greek *agon*, meaning contest, from which "agony" also derives. Entomologists sometimes refer to agonism as "antagonism.") Agonistic behavior is usually viewed as a continuum that ranges from pure fighting or attack at one extreme, through ambivalent behavior and signaling, to pure flight or escape at the other extreme. The diversity of agonistic behavior is wide, and an implicit classification in much of the ethological literature recognizes: (1) personal space or individual distance, in which an animal attempts to exclude other animals from some volume of space around itself; (2) resource priority, in which an animal attempts to exclude others from the space around some discrete resources, such as a food-item or nest-site; (3) territory, in which an individual attempts to exclude others from a relatively large environmental space that may contain several resources, such as food, hiding places, nest-site, *etc.*; and (4) dominance, in which an animal attempts to establish a general priority over other individuals without defending a particular spatial locus.

It is a convenient oversimplification to divide communicational signals of agonism into those of threat and appeasement. In a general sense, a *threat signal* has as its semantic referent conditional probabilities of overt attack. The probabilities are conditional upon the behavior of the receiver. For example, during establishment of a territorial boundary, a threat signal could indicate that the sender will probably attack if the receiver comes closer, may attack if the receiver remains, and probably will not attack if the receiver retreats. Regardless of which the receiver does, it may also reply with a signal. The original sender's threat may not have effected any spatial alteration in the relationships of the two animals, but the sender has benefited by gaining information about his opponent's own probabilities of attack. Often, a dispute may be settled by an exchange of signals without either animal risking injury or death through overt combat.

Appeasement signals appear to be given when an animal attempts to remain in some place without challenging the dominance of another animal. Appeasement signals appear to have as their referents low probabilities of overt attack. Threat seems to be more widespread, but appeasment signals are important in dominance hierarchies and in early phases of sexual behavior.

Reproductive behavior includes a very broad spectrum of activities which may be decomposed into temporally overlapping phases for purposes of discussion. In the first place, it is necessary to attract a potential mate of the same species and opposite sex, the process of *mate-attraction*. (In synchronously hermaphroditic species, early specification of sex is of course unnecessary.) Of potential mates, one or more animals must be selected for ultimate spawning, copulation or equivalent behavior of fertilization, and the selection process may be called *mate-choice*. In some species there is a protracted bond between the potential parents, and hence mate-choice in these species is often referred to as "pair-formation," "bonding," or similar terms. The third major phase is fertilization

(spawning, copulation, *etc.*) and all the activating and
synchronizing activities that lead up to it. Ironically,
there appears to be no wide-spread ethological term designating these activities; "courtship" usually includes
mate-attraction and mate-choice whereas "precopulatory
behavior" usually refers only to actions immediately before copulation. For lack of a better term, I shall call
this class simply *sexual behavior*. Finally, I shall combine all activities relating to the rearing of offspring
as *parental care*. Such activities could be subdivided into sheltering the young, feeding them, protecting them
from predators, *etc*. However, parental care is virtually
absent in many species, it is extremely diverse where it
occurs, and it raises only a few general problems of signal-design, so relatively little special attention is accorded it in this discussion.

The third major functional category of social behavior that involves communication is *cooperative behavior* in
a very general sense. Under this rubric I discuss all
communication that is not patently agonistic and that is
not restricted to potential mates, mates, or parents and
their offspring. Three subcategories may be distinguished:
social maintenance activities, such as allogrooming in mammals; *coordinated group movements*; and *special group activities*, such as cooperative foraging and anti-predator
behavior. Many of these may be elements of reproductive
behavior, as when one animal warns its mate about the
presence of a predator, but none of these activities is
restricted to reproduction. Indeed, if current views of
social behavior be correct (J.L. Brown, 1975; Wilson, 1975),
many of these cooperative activities may be expected as
particularly common elements of parental care. Cooperative
behavior within the family is the most likely evolutionary
origin of more general social behavior in extended groups
of genetically related animals.

Table 8-IV (next page) summarizes the classificatory
scheme used to discuss pragmatic aspects of optical signals. Each subcategory recognized could readily be further subdivided, but the present scheme is adequate for
purposes of analysis.

threat

The general sematic referent of threat signals is con-

Table 8-IV

A Classification of Social Behavior

Behavioral Category	Subcategories
agonistic behavior	threat appeasement
reproductive behavior	mate-attraction mate-choice sexual behavior parental care
cooperative behavior	social maintenance activities coordinated group movements special cooperative activities

ditional probability of overt attack. The probabilities depend upon the subsequent actions of the receiver: its spatial movements, returning signals and so forth. Threat includes long-distance signals such as those given by a songbird establishing a territory and at the other extreme signals given when animals are within physical striking distance of one another. In the former case, threat signals need to optimize conspicuousness, often in a complex environment, so that principles of *ch 7* will apply. When the function of threat is to exclude only conspecifics, as in the case with territory in many songbirds, threat signals may be species-specific. However, many are not. My own notes on more than a dozen emberizines show head-forward threat to be highly similar among species (see figure 23 in Hailman, 1977a). Furthermore, many agonistic encounters are interspecific, in which case one expects convergence of signals to a common type (Cody, 1969).

The most general characteristic expected of threat signals is *iconicity for attack*. Most threat signals appear to have evolved from incipient movements of attack (Hailman, 1977a) and resemble fighting postures, movements and orientations. Often, the major benefit the sender gains from a threat signal is a reply from the receiver that allows the sender to judge the receiver's own probability of attack (Maynard Smith, 1974; Parker, 1974).

When the sender can also effect retreat by the receiver, the benefits are greater. One therefore expects threat signals to be constructed so as to exaggerate the invincibility of the sender, and this exaggeration seems to have taken at least two forms: display of weaponry and appearance of large size.

Many threat signals *display weapons* used in fighting (*e.g.*, N. Tinbergen, 1959; Hailman, 1977a) as part of the icon (*figs 8-5a* and *8-5b*). For example, Miller (1975) re-

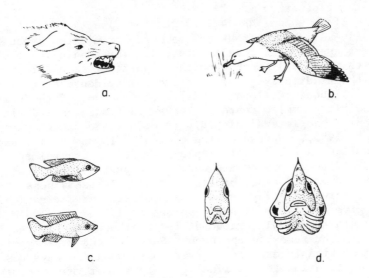

a. b.

c. d.

Fig 8-5. Elements of threat-signals. The expression of the dog (a) and pecking into and pulling at grass by the herring gull (b) are iconic signals predicting fighting. The signals display the primary weaponry of the dog (teeth) and the gull (wings and bill). Threat displays in fishes may make the animal look larger, as in the lateral display of Tilapia natalensis *(c, lower) compared with resting (upper). The frontal display of* Cichlasoma meeki *(d, right) also shows an increase in size over the normal posture (left). Dog after Darwin (1871), gull after Tinbergen (1951) and fishes after Baerends and Baerends-van Roon (1950).*

ports that nearly 80% of threats by male walruses involve display of the tusks. Tusk-size is correlated with body-

size, but when the effect of the latter is removed statistically, there is still a tendency for males with the
largest tusks to win supplanting encounters. More interesting is the fact that animals have evolved structures
and color-patterns that mimic weapons; for example, the
manes and ear-markings of ungulates with horns (see figure
22 in Hailman, 1977a). The "automimetic" elements of
threat display probably do not actually deceive opponents
into thinking that the sender possesses extra weapons;
rather, the icon of weapon-display is visually enhanced
by the redundancy (ch 7).

     If an animal *appears larger* (or closer) than it ac-
tually is, the opponent may be more intimidated because
overt approach is itself iconically threatening. Ethologists repeatedly emphasize the importance of apparent size
in threat displays, but experimental evidence on this point
is ambiguous (Hailman, 1977a). Perhaps one of the best ostensible examples concerns threat in fishes, particularly
frontal and lateral displays (Baerends and Baerends-van
Roon, 1950). Spreading the operculum in the former and
the fins in the latter display creates a larger image
than if these components were not a part of the signal
(*figs 8-5c* and *8-5d*, previous page). It seems possible
that threat displays could also be structured dynamically
to give the impression of approach. If the cichlid's o-
percula are slowly extended, for instance, the increasing
size might resemble approach. Such *looming* aspects of
threat signals are easily overlooked when one pays descrip-
tive attention to the final posture rather than dynamic
movement in signaling.

     It is also possible that coloration may be utilized
to increase apparent size. A light object against a dark
background looks larger than the reversed polarity because
of irradiance (ch 5). Leonardo da Vinci recognized that
white clothing made persons appear larger than reality
(Minnaert, 1954: 105), and it is possible that animals use
the same principle. For example, one threat posture of
the white-throated sparrow consists of tilting its bill
upward to display the throat, which may appear larger than
normal because of irradiance. Many other emberizines have
less dramatic white throats, but a few species have black
throats. The ruff also provides a counter-example. The
aggressive resident males of this shorebird species maintain territories in the lek and have large, darkly feathered
ruffs, whereas the appeasing satellite males have white

ruffs (Hogan-Warburg, 1966; Rhijn, 1974). The coloration
may have to do with conspicuousness in their particular
habitat.

Threat signals may be very diverse. There may be different signals to indicate different conditional probabilities of attack for different circumstances. One aspect
of signals that might be expected is syntactic *quantification* of the level of probability. As mentioned previously, J.L. Brown's (1964) study of crest-raising in the agonistic encounters of the Steller's jay suggests that such
quantification is being communicated by the angle of the
crest.

appeasement

There appear to be at least two distinct classes of
signals given by an animal attempting to remain near another without antagonizing it: antithetical and iconic
appeasements. In *antithetical appeasement* the sender gives
signals "opposite" to threat signals: it faces away from
the opponent, hides weapons, looks as small and distant as
possible and so on (*fig 8-6a*; see also figures 23 and 24
in Hailman, 1977a). Indeed, simply avoiding eye-contact

a.   b.

*Fig 8-6. Signals of appeasement. Elephants (a) demonstrate
Darwinian antithesis in that the most threatening posture
(left) extends the trunk, points the tusks forward, and ro-
tates the ears forward; antithetical elements (right) include pulling the trunk down and backward, lowering the
tusks and appressing the ears against the head. In the
avocet (b) pseudosleeping in agonistic situations is an example of iconic appeasement, which also hides the principal
weapon (the bill). Elephants after Kühme (1963), who sepa-
rates many intermediate signals, and avocet after Makkink
(1936).*

is an important appeasement signal in canids and felids
(Kleiman and Eisenberg, 1973), as well as in many primates
and other animals. Such antithetical signals are often
iconic for fleeing as well.

Less obvious are more specific *iconic appeasement* signals that resemble behavior other than agonism. For example, many primate species present the anogenital region
when approached by a dominant individual, and in some baboons the entire area is permanently swollen and reddened
to resemble the posterior of a female in estrous (see illustrations in Wickler, 1968). Such icons are mimetic,
but probably do not work by true deception in the sense of
ch 6. In early evolutionary stages an appeasing sender
might well have deceived the dominant individual, but it
seems reasonable to conclude that the appeasement signal is
now recognized as a signal, standing for a referent such
as "I will behave more like an estrous female than a rival
male." Similarly, the dominant male's perfunctory mounting
that often follows such presenting is iconic in the same
sense of resembling non-agonistic behavior. These signals
are sometimes called "remotivating" signals, but there seems
to be little firm evidence that the animals are "remotivated."

In general, one may expect that any non-agonistic behavioral pattern may be mimicked by an iconic appeasement
signal. The classical example is pseudosleeping in fighting avocets (Makkink, 1936), which in many ways looks like
sleeping and hence may signal to the opponent that the sender is not likely to attack or move from the spot *(fig 8-6b,*
previous page). In this case, it is even clearer that the
opponent is not "remotivated" to engage in some particular,
irrelevant behavior.

Lorenz (1952: 186) proposed a type of appeasement signal based on showing the most vulnerable part of the body
to the opponent. The signal was supposed to be symbolic
for a referent such as "kill me if you wish as I can no
longer resist." His idea was based on a signal of canids
where one agonist turns his head away to expose his neck,
and hence the jugular vein. Lorenz apparently misinter-
preted the observations on wolves by Schenkel (1948, *pers.
comm.*); in fact, Schenkel claims that the signal is given
by the dominant animal, as a test to see if it has won. If
the opponent resists the chance to attack the displayer's
most vulnerable area, then the dominant animal realizes
that the contest is over and his dominance is complete. I
might add that in my college youth I put Lorenz's theory

Table 8-V

Optical Characteristics of Agonistic Signals

Signal-type	Optical Principle	Specific Characteristics
threat	conspicuousness	see *ch 7*
	attack iconicity	
	weaponry display	mimetic redundancy
	appearance of largeness, near- ness or approach	spreading of appendages piloerection irradiance of white coloration
appeasement	threat-antithesis	hide weapons appear small be concealing
	non-agonistic iconicity	feigning (*e.g.*, pseudo- sleeping)
both	individual dif- ferences	differences in elements of or near basic ago- nistic signals

to empirical test by throwing down my snowballs when cornered, and opening my arms defenselessly. The result was
a snowball in my face.

Finally, in many agonistic encounters, especially
those related to dominance hierarchies, *individual recognition* is important. Because agonistic signals are often
encephalized for reasons discussed previously, one might
expect variations in signals about the head region to be
important cues to individual identity. For example, in
domestic fowl, differences in the comb and wattle appear
to provide information about individual identity (H.L.
Marks *et al.*, 1960).

Some of the characteristics expected in agonistic
signals are summarized in *table 8-V*.

mate-attraction

The two major requirements for a potential mate are
that it be of the same species and the opposite sex (except in synchronous hermaphrodites). Mating with another
species wastes time and gametes because such matings often
produce no offspring, sterile offspring, less viable offspring or offspring that are at a competitive disadvantage
ecologically (see Mayr, 1963). There is therefore a strong
evolutionary selection for making a conspecific mate-choice,
and studies reveal consistent differences in optical signals among closely related species (*e.g.*, Hunsaker, 1962).
The necessity for correct attraction of potential
mates has often been used to explain dramatic differences
in display coloration between male and female, and among
males of different species. However, it appears that the
explanation has been employed uncritically and overworked.
Many closely related species whose breeding ranges overlap
have highly similar signals, yet rarely interbreed. For
example, males of the common and Barrow's goldeneye ducks
differ only slightly in display markings (see figure 9 in
Hailman, 1977a), and the virtually identical black-capped
and Carolina chickadees differ primarily in only one type
of vocalization. Conversely, the wood duck and mandarin
duck have no potentially interbreeding species within
their separate ranges, yet the males have highly elaborate
display plumages (Dilger and Johnsgard, 1959). Sexual dimorphism is just as difficult to explain simply. Male and
female black ducks, for example, are sexually monomorphic,
yet the very closely related mallard shows extreme sexual
dimorphism in coloration.
Sibley (1957) suggested that sympatic, closely related
avian species that are polygamous and have short pair-bonds
should show the greatest sexual dimorphism and species-distinctiveness in courtship signals. In such species males
would be highly competitive for the females (part of the
notion of sexual selection). However, Dilger and Johnsgard
(1959) pointed out that monogamous species with long-term
pair-bonds might suffer even more strongly from incorrect
pairing. They point to species such as parrots that show
elaborate species-distinctiveness, and in some cases sexual
dimorphism, yet have long pair-bonds. And mallard and
black ducks show the same general reproductive patterns,
yet one is dimorphic and the other is not.
T.H. Hamilton (1961) suggested that it is not the

length of the pair-bond, but rather the duration of courtship that correlates with sexual dimorphism. Migratory
species with short breeding seasons must arrive in spring
and immediately establish territories and secure mates if
any offspring are to be produced. In species with rapid
courtship, sexual and specific distinctiveness must be
emphasized to prevent costly mistakes. Yet the mallard
and black duck, to cite one example, are both migratory.

T.H. Hamilton and Barth (1962) extended the idea to
seasonal dimorphism, where the male assumes female-like
plumage during the non-breeding season, not only to avoid
predation through concealing coloration (*ch 6*), but also
to reduce hostility in wintering flocks (also see Moynihan, 1960). Their evidence for the latter point is scanty
and may not be correct. The male Baltimore oriole, which
they cite as principally solitary on the wintering grounds,
retains its bright orange plumage, whereas the male or-
chard oriole, which winters in social groups, is cited as
molting into a dull, female-like plumage. However, handbooks for areas within the wintering range of this latter
species illustrate and discuss only the bright plumage of
the male (*e.g.*, Peterson and Chalif, 1973; de Schauensee,
1970).

Willson and von Neumann (1972) surveyed the "colorfulness" and sexual dimorphism in birds in both the Old World
and New World and found certain statistically reliable
differences among regions. North American temperate and
South American tropical species did not differ in sexual
dimorphism (39% possess it), but these avifaunas show higher incidence than either Europe (32%) or South American
temperate birds (26%). "Colorfulness" had a different
geographic pattern. It was highest in Neotropical birds
(32%), less common in South (27%) and North (25%) American
temperate avifaunas, and even less common in Europe (10%).
They offer no new hypotheses to account for these differences. It might be that "colorfulness" is related to conspicuousness in the dense vegetation of the tropics (T.
Johnston, *pers. comm.*). The requirement would be less
stringent in more open temperate forests, and even less important in Europe, which has been largely deforested for
a very long time.

In sum, no one has established a hypothesis that predicts with reasonable accuracy the degree of colorfulness,
sexual dimorphism or specific distinctiveness in courtship
plumage of birds. Fishes present the same problem, as var-

ious species are brightly colored or dull, similar to one
another or strikingly different, sexually monomorphic or
dimorphic, *etc.* Part of the problem in analyzing fishes
is that many change color readily *(ch 4)* and detailed
studies are required to describe all the colors shown by
a species under different behavioral conditions. Among
insects there is even another problem: UV-sensitivity.
The black and white butterfly *Eroessa chilensis,* for example, has orange wing-tips in both sexes and thus appears
monomorphic. However, Eisner *et al.* (1969) discovered
that only the male's wing-tips reflect ultraviolet, so
the species is actually sexually dimorphic.

As noted, the important aspects of initial mate-attraction that affect characteristics of optical signals
are *species-specificity* and *sex-specificity.* Concerning
the former, one expects only the characteristic that species should be discriminable on some basis (not necessarily optical); there is no clear expectation for particular
characteristics when optical signals are used. Concerning recognition of sex, two factors in the design of optical signals may be expected: secondary sexual characteristics and sexual antithesis.

First, any features that are characteristic of one
sex due to non-signal functions, such as weaponry of males,
may be exaggerated optically or prominently displayed to
assist sexual recognition. Such elements are called *secondary sexual characteristics,* and the structures and color-
patterns that automimic weapons (see *threat,* above) may be
further selected to act in mate-attraction *(fig 8-7a).* The
other expectation is that sexes will be as "opposite" as
possible in signaling: they will exhibit *sexual antithesis*
(fig 8-7b). The difference between male and female may
be virtually permanent, as in the coloration of cardinals,
or virtually instantaneous. For example, Keenleyside (1972)
says of the pomacentrid fish *Abudefduf zonatus* that the
male's "pattern appears suddenly as the male begins court-
ing and fades quickly when the female leaves the male's
nest area."

Finally, behavior is organized in some animals such
that a male establishes a territory from which he simultaneously attempts to exclude other males and attract females. Signals that appear to serve both functions are
called *advertisement,* a common example being bird song
with display of brightly colored plumage. Advertising
signals may be expected to maximize conspicuousness *(ch 7).*

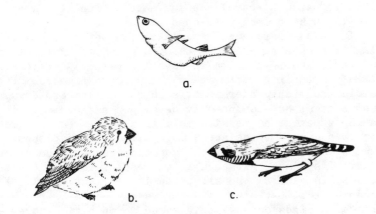

Fig 8-7. Sex-specific signals of mate-attraction. The female three-spined stickleback (a) when ready for repro-duction can be recognized by her shape, the ventrum dis-teneded due to accumulation of eggs. Her posture displays the shape prominently before the male. The female zebra finch (b) lacks the bright coloration of the male (c). The sexes are shown in antithetical postures: the male in a sleeked, head-forward threat (cf., fig 8-5) and the fe-male in a more upright, fluffed posture (cf., fig 8-6). Antithetical and iconic appeasement by females are ele-ments of sexual behavior (see fig 8-9, below). Stickle-back after Tinbergen (1951) and zebra finches after D. Morris (1954).

Extrinsic signals may also be used to advertise: Linsemair (1967) reports that the sand pyramids built by ghost crabs simultaneously repulse other males and attract females.

mate-choice

Darwin (1871) raised a major question about reproduc-tive behavior that has been accorded too little experimen-tal attention: by what characteristics should an animal choose a mate in order to maximize reproductive success? There appear to be at least three major classes of criteria that would be useful in mate-choice: the genetic comple-ment of the mate, the mate's parental abilities (when in-

volved in parental care) and the mate's fidelity (when
there is to be a long pair-bond).

There has been considerable confusion as to how one
animal assesses superior genes of a potential mate. The
classical idea is that males reveal superior *genetic complement* in their competitive fighting over females. For
example, courtship adornment, such as antlers and horns
of ungulates, have supposedly developed as means of displaying superior aggressive abilities *(fig 8-8a)*. A female is supposed to select a male with well developed
weapons because these reflect superior genes to be passed
to her offspring. Although it may work this way in some
species, in general the situation is not that simple. For
instance, the most aggressive males may not make the best
parents in birds because they spend so much time in agonistic activities that they neglect the needs of their offspring and mate (Hutchinson and MacArthur, 1959).

It is not impossible that large horns and antlers in
ungulates, and similar developments in males of other animals, reveal important aspects of ecological acumen in
their possessors. Perhaps a male with large antlers demonstrates to the female that he is proficient in extracting critical nutrients from the enviroment. If such traits
are genetically controlled, the female would do well to
insure that her offspring receive the beneficial genes for
good foraging. I am aware that this is not a traditional
explanation of secondary sexual characteristics, but it is
a hypothesis worth closer attention. To cite a related
speculation, S. Robinson *(pers. comm.)* has found that male
house sparrows bill-wipe more frequently in the presence
of a female than they do in the presence of another male,
or when alone. He suggests that this subtle signal may
have arisen in evolution if females tended to choose mates
who were proficient in feeding, and hence wiped their bills
frequently.

When the male participates in parental care, these
same aspects of morphological and behavioral signals may
be further enhanced by at least two other mechanisms. Either the male might promote the offspring's learning how
to feed optimally, or else the male may be a good provider
of food for the offpsring (either by bringing food to the
offpsring or leading the family to food). Thus, genetically
determined ecological acumen grades into predictors of
the mate's *parental abilities*, the second category of attributes that may be signaled in communication leading to

mate-choice.

   If success as a parent is correlated with age, mate-choice may be based partly on optical indicators of age (S. Witkin, *pers. comm.*). For example, the size of antlers and horns in ungulates is known to be a function of age in several species, thus providing another possible value of such structures as optical signals. Studies of individuals in primate groups, particularly great apes such as the gorilla and chimpanzee, show silver hairs to be age-indicators. In the human primate there is great variation in the age of graying (my mother's hair was snow-white in her teens, although I kept color a bit longer), so age-indicators may not be very accurate. In fishes and invertebrates, and probably to some extent in reptiles and amphibians as well, size alone is correlated with age and might play a role in mate-choice.

   Ornithologists have long recognized that female songbirds may choose a mate more by the quality of territory he holds than by his own physical characteristics. She is optimizing her reproductive strategy, and it may be more adaptive to become a second wife on a good territory than the sole mate on a poor one (Orians, 1969). The territory is one aspect of predicting the ability of the male to be a good parent. Other aspects are commonly noted among birds. For example, grebes build a nest-like platform for copulation (Huxley, 1914). Because they cannot copulate in the water in the fashion of ducks, the platform is a functional structure in a straightforward sense, but its resemblance to nests is probably not accidental. It seems reasonable to assume that by constructing the platform, the potential parents communicate to one another something about their abilities to provide a good nest for ultimate eggs and young. Nest-like structures are important in the courtship of many birds (*e.g.*, Verner, 1964), and potential parents often construct several nests within the territory before actual egg-laying. These have often been viewed as simple failures and abandonment, but it seems possible that in some cases they involve elaborate testing of the mate's parental abilities. The nests are extrinsic signals: iconic predictors of the ultimate nest to be built.

   Many of the elaborate courtship displays of birds have been attributed to simple symbolic actions of unspecified origin. I believe that these should be scrutinized for predictors of parental abilities. For example, returning to Huxley's (1914) great crested grebes, the pair

engages in an elaborate "penguin dance" in which both birds
dive to the bottom and emerge with a bill full of vegetation like that used in constructions of the nest (fig 8-8b).

Fig 8-8. Signals used in mate-choice. The male's rack
of antlers in the moose (a, left) not only prepares him
for fighting rivals but may also indicate to the female
his possession of desirable genes, such as those that produce good foraging abilities. The "penguin dance" of the
great crested grebe (b, right) involves showing nest-like
materials gathered underwater, and is thus an icon that
demonstrates potential nesting abilities of the birds
(after Huxley, 1914).

Wilson's (1972) interpretation that "the collection and
presentation of the waterweeds may have evolved from displacement nesting behavior initially produced by the con-
flict between hostility and sexuality" seems to me unnecessary. It is more parsimonious to assume that evolution
favored grebes that could select their mates at least partially on the birds' abilities to gather nesting material
successfully. Once receivers were cuing on nest-material
gathering as part of mate-choice, evolution favored making the gathering visually obvious, and highly ritualized
optical signals evolved. The more ethology has learned
about principles of communication, the less it has had to
rely on explanations such as "displacement activities"
(see also Hailman, 1977a). Eibl-Eibesfeldt (1970: 127)
describes taking food-like items away from Galapagos
flightless cormorants, which usually present the materials

to the mate at the nest-site. Without the iconic gift, the bird at the nest rejects the returning mate. More experiments need to be done to establish the importance of extrinsic signals in courtship communication.

If both animals participate in courtship-building, the activities allow each to judge how well the other will coordinate activities so that would-be parents can work as a team in reproductive activities. Individuals of a species differ in their daily rhythmic patterns, and some kinds of courtship activities may test the abilities of the two sexes to coordinate their daily schedules in useful ways. It seems likely that courtship-feeding (Lack, 1940) is another example of how courtship activities help one mate to judge parental qualities of the other. If a female bird incubates eggs during the day, she may later have to depend upon the male to bring her food. If the male helps feed the young, that provides another reason to assess his food-bringing abilities before a final commitment to reproduction (Nisbet, 1973). I believe that animals learn a great deal more about the parental abilities of their mates through courtship activities than has been realized. These are not simple, arbitrarily symbolic actions; many are iconic predictors of parental abilities.

Finally, when the mate's presence after copulation is important to the success of the reproductive effort, mate-choice by the partner may focus on predictors of the *mate's fidelity*. The female often has a larger investment in the reproductive effort, since she produces the nutritional-bearing gametes at relatively greater physiological cost than production of male sperm. Males may evolve an optimizing strategy of attempting to inseminate many females, so polygyny is to be expected more frequently than polyandry. McKinney (1975) relates the use of true deception in male green-winged teals, who assume the sleeping posture and induce their mates to sleep; then the male sneaks off to attempt copulation with some other female. The subject of mating systems is too involved to treat in this book, but one can predict that one aspect of optical signals relating to mate-choice may concern attempts of the female to test her mate's fidelity. A typical example was mentioned in connection with indexic aspects of semantics, above: female mallards attempt to incite their consort males when approached by an intruding male *(fig 8-2a)*.

It follows from aspects of choosing an individual animal for a mate that they may also be signals for *individ-*

ual recognition. In many cases, variation among individuals due to other causes may provide sufficient cues to individual recognition. In some cases, however, individual variation may be selected for its signal value, as in individual differences in display movements of the chuckwalla (Berry, 1974).

Finally, copulation itself (or functional equivalents of fertilization processes) may play an important role in mate-choice. Non-fertilizing copulations may take place in early phases of courtship, where such acts serve as iconic predictors of later sexual behavior and may also indicate the mate's fidelity. In some cases, elaborate post-copulatory displays have evolved, and although I believe these to relate primarily to mate-choice, they also function in maintenance of pair-bonds, and so are discussed most naturally in the section that follows.

sexual behavior

I have included under the term "sexual behavior" all those activities that take place between the time of mate-choice and fertilization. These activities are very diverse among animals, depending on the life-history strategies of reproduction, and hence relatively few generalizations are possible. Some primary considerations appear to be: overcoming the mate's aggressiveness, stimulating and synchronizing sexual development of the pair, and actual fertilization (copulation, spawning, _etc._).

Because males either fight for possession of females or defend breeding territories in many species, overcoming male aggressiveness may be an important factor in the female's reproductive success. In some animals, the male may have to overcome aggressiveness of the female. This seems to be particularly the case with predatory species, such as the balloon flies mentioned previously, some spiders, raptorial birds, solitary carnivores and other animals. The general problem is therefore _overcoming the mate's aggressiveness_, and in most species the more aggressive sex will be the male.

Meyerriecks (1960) notes that the female green heron just keeps returning to the male of her choice after being driven off, and so eventually is accepted more because of doggedness than any special communication signals. Of course, this behavior may also be part of the male's test-

ing of the fidelity of the female. In many species, however, females show special iconic appeasement, a common
form being signals that resemble begging in young animals
(fig 8-9). The male, in turn, often feeds the female (Lack,

*Fig 8-9. Signals used in sexual behavior. The female may
decrease the male's hostility by assuming antithetical
appeasement postures (see fig 8-7b) or both mates may engage in behavior that hides the weapons and is appeasing,
as in the laughing gull's facing-away (left, after Beer,
1975). The female may also engage in iconic appeasement
(see also fig 8-6b), as in the kittiwake (right), where
she assumes a begging posture like that of young birds
(after Tinbergen, 1959). The begging posture, accompanied
by crouching that is iconic for the copulatory stance, is
also a precopulatory signal in female gulls.*

1940), as mentioned in the previous section. Ornithologists have frequently interpreted the dull coloration of
females as being an adaptation for concealment, especially
in species where the female incubates during the day. T.H.
Hamilton (1961) suggested that in sexually dimorphic species such coloration also acts as appeasement, being antithetical to the bright plumage of the male *(figs 8-7b* and
8-7c, p. 283). Therefore, both *iconic appeasement* and *antithetical appeasement* occur in sexual behavior.
      Behaviorally antithetical appeasement also occurs in
sexual behavior. Emory (1976) reports that simple orientation toward another baboon is threatening; semantically
the signal is indexic, syntactically it is addressing a
specific receiver. Gelada males apparently consciously
orient so that they do *not* face certain females, the body-
orientation being a specific form of sexual appeasement.
Head-flagging between male and female gulls *(fig 8-9)* is

similarly a cutoff signal thought to reduce agonism between mates (N. Tinbergen, 1959; see also Beer, 1975).

It is not clear that any special characteristics of
signals that *stimulate* and *synchronize* the reproductive
development of the mates are to be expected. The subject
seems worthy of specific consideration. Immediate *preparation for fertilization*, however, is expected to involve
icons of copulation itself, such as lordosis postures of
females. Display of the erect penis in many primates
(*e.g.*, chimpanzee: van Lawick-Goodall, 1968) is obviously
iconic. In some species it has been incorporated into
earlier phases of reproductive behavior, where it is used
in identification of sex in mate-attraction, and in some
primates the penis and genital area have become brightly
colored to increase conspicuousness and perhaps enhance
shape (see discussion previously). In sexually monomorphic
species such as the starling (Hailman, 1958) the female
may mount the male as well as *vice versa*. It is not impossible that reverse mountings in birds without intro-
mittant organs can lead to insemination, but it may be that
these mountings are pre-copulatory iconic signals.

In many cases, one may expect indexic signals to facilitate fertilization. Thus, in the creeping-through behavior of the stickleback mentioned previously, the male
shows the spawning site to the female, and in the display
of the courtship paddle of certain fishes (*fig 8-2c*, p. 253)
the male indexes the spatial position of the female for
internal fertilization. A third example from fishes, that
illustrated in *fig 8-2d* where the female mouth-brooding
cichlid picks up sperm to fertilize the eggs in her mouth,
emphasizes the diversity of indexic signals in fertilization.

Synchronous hermaphrodites are not common among higher
animals, although in certain groups of fishes an individual
may be one functional sex for an early period in life and
the other sex later. If cross-fertilization is to occur
in synchronous hermaphrodites, either the two individuals
must release opposite gametes, or else at least one of them
must release both sperm and eggs. Clark (1959) studied
optical signaling of the serranid *Serranellus subligarius*
and found that one fish from a group often began swimming
in an S-curving pattern, blanched in color (losing the
dark bands on the body and black spot at the base of the
dorsal fin) and changed body shape. If another fish responded similarly, it was chased by the first until it

returned to the banded pattern. The blanched fish appears
more likely to release eggs and the banded fish sperm,
but in aquaria one individual may release both gametes
and self-fertilization is possible, so the situation is
not simple. At any rate, it appears as if a form of sex-
recognition is involved in immediate pre-spawning behavior
of synchronous hermaphrodites.

Many species have elaborate *post*-copulatory display
behavior, which has been little analyzed by ethologists.
For example, the male mallard dismounts, assumes a particular bridling posture and whistles, then nod-swims in
a broad arc in front of the female. It seems likely that
these post-copulatory displays have evolved as part of
mate-choice (above). Mallards copulate for months before
nesting (Weidmann, 1956), and unless there is long-term
sperm-storage, these copulations probably function more
in mate-choice than in functional fertilization (McKinney,
1975). I believe the male's display is to announce to the
female successful intromission, since in copulatory attempts
that clearly do not result in cloacal contact, the male
omits post-copulatory display (Hailman, *in prep.*). After
bridling, the male swims in front of the female so that
she can see again the individual with whom she has successfully copulated. The male may also be demonstrating
his intentions of fidelity (McKinney, 1975).

parental care

Parental care involves many activities, but the primary communication occurs between the two mates (when both
are present) and between parents and their offspring.
Parental care is highly diverse in animals, many species
lacking it entirely, so relatively few generalizations are
possible. Many aspects of parental care involve more general cooperative behavior that may be shown other animals
as well, and hence are treated in the subsequent sections.
A few comments about special communication relating to
parental care may be added here.

When there is a special site for eggs or young, the
mates may have to agree upon the site. Depending upon the
nature of the site, which is very diverse among species,
special indexical signals may be involved. For example,
the dorsal display coloration of hole-nesting birds was
mentioned previously (von Haartman, 1957). When caring
for eggs or young, the parents may have to synchronize

their activities, and in some cases of birds, special
"nest-relief" ceremonies have evolved, although their optical properties have not been well analyzed. Many may
be simple icons for incubation and agonistic behavior
(Beer, 1961; see also Armstrong, 1947).

Feeding is a primary interaction between parents and
offspring, with the young animals evolving special signals
to indicate that they are hungry. For example, most young
songbirds gape to their parents, and the gapes are decor-
ated with patterns of color that enhance conspicuousness
and in many cases may be species-specific. Friedmann (1960)
pointed out that parasitic weaverbirds, which lay their
eggs in nests of other species, have young with gape-sig-
nals resembling those of the host species. In some cases
it has been suggested that gape signals are icons resembling other things, but the whole area of investigation
requires further work. For example, Hingston (1933) suggested that the rows of white conical projections in the
gapes of nestling bearded tits were mimics of palatal
teeth of reptiles to frighten off nest predators.

The parents, in turn, may have special signals concerning feeding of the young. In gulls, the conspicuous
red markings on the beak serve to test the hunger state
of the chicks, which beg by pecking if they require food.
The parent responds to the pecking by regurgitating partially digested food, and then indexes its location by
pointing to the food with its bill. Chicks pecking at
the bill-tip strike the food for the first time, and thus
begin to learn its visual characteristics (Hailman, 1967a).
In mammals, the female may signal willingness to nurse offspring by very simple visual signals, such as lying on her
side to expose her teats or straddling her young. Such behavior is iconic and in some cases indexic, and no doubt
the coloration of teats in some mammals serves primarily
an indexic function in guiding the young.

Table 8-VI summarizes some of the attributes of signaling in reproductive behavior discussed above. It is
intended to be suggestive of the kinds of variables that
should be scrutinized in order to understand optical signals, rather than an inclusive classification of all relevant variables.

social maintenance activities

It is easy to see how one animal may derive benefit

Table 8-VI

Optical Characteristics of Reproductive Signals

Behavior	Signal-type or Function	Specific Characteristics
mate-attraction	species-specificity	?
	sex-specificity	sexual icons sexual antithesis
mate-choice	mate's genetic complement aggressive success	threat signals*
	ecological success	morphological development behavioral icons
	mate's parental abilities	extrinsic icons (nests, *etc.*) behavioral icons
	mate's fidelity	testing behavior
	individual recognition	post-copulatory display
sexual behavior	appeasement	threat-antithesis* non-agonistic iconicity*
	sexual development and stimulation	?
	coordination of fertilization	copulatory icons indexic guides
parental care	parent-parent signals	site-indexes duty-icons
	parent-offspring signals	feeding icons and indexes cooperative behavior#

*See table 8-V; #See table 8-VII.

from allo-maintenance activities by another. For example,
it is physically impossible for a bird to preen its head-
feathers with its bill. The bird may be able to effect
the same result by other means, such as scratching with
its toes, but one solution is to have another bird preen
its head. Or, if grooming in monkeys serves partly to
find and remove ectoparasites in the fur, a monkey that
cannot see its own back or reach it conveniently benefits
from another monkey's allogrooming.

Social maintenance activities are not, however, always that straightforward. In many cases they have become incorporated into agonistic and reproductive behavi-
or: subordinates grooming dominants, mates grooming one-
another, parents grooming their offspring, *etc*. (see Harrison, 1965; Sparks, 1965). Indeed, the behavior may
cease entirely to serve as functional maintenance activities and become instead signals of various kinds. It is
not my intent here to discuss these intricacies of social
maintenance activities, but rather to ask what expectations
one may hold about the optical signals that accompany them.

One expectation is simply that of indexing, usually
by orientation, the area the sender wishes to have groomed
and addressing the invitation to a particular individual
(fig 8-10). There are often iconic signals of maintenance
itself, such as ruffling the feathers, which is done prior
to some kinds of self-preening in birds. Such signals have
also evolved into deceptive mechanisms of interspecific
communication, as when the brown-headed cowbird female
(which lays her eggs in the nests of other species) solicits
allopreening from the host birds by displaying ruffled head-
feathers (Selander and La Rue, 1961).

coordinated group movement

Coordinated movements of whole groups of animals are
common. Movements may be short, as in daily troop wander-
ings of baboons, or lengthy, as in transcontinental migrations of some birds. They may be periodic or virtually
constant, as in the schooling of many fishes. The benefits
of group movement may relate to specific activities, such
as cooperative foraging or predator-defense (discussed
below), but whatever the benefits there must be signals
of some sort that coordinate and integrate the movements
of the component animals. In most cases the signals will
be optical (but see Pitcher *et al.*, 1976).

Species-specificity is not a universal trait of signals that coordinate group movements because mixed-species
groups are well known (*e.g.*, Moynihan, 1968). In such
cases, there may be convergence of coloration or other attributes that promote recognition, as in the black coloration of huge interspecific flocks of birds during the winter in southern United States. (Some groups consist of
literally millions of red-winged blackbirds, rusty blackbirds, common grackles, brown-headed cowbirds, starlings
and other species.)

One trait expected in signals that promote group movements is *iconicity for movement* itself. Incipient movements of locomotion are common in birds (Daanje, 1950) and
these are ritualized to various degrees for use as signals
promoting synchronous action (see Andrew, 1956 for a discussion of avian tail-flicking as a typical example). In
other cases, the iconic nature of locomotion-promoting signals is not immediately obvious. For example, the rapid,
jerky uptilting of the mallard's bill, which one of my
students dubbed "lid-flipping," can be seen upon scrutiny
to resemble the head movement that is part of take-off from
the water. However, lid-flipping is not given invariably
before take-off nor do mallards always take off after lid-
flipping. The action is a deliberate signal to recruit
social movement: if other birds begin lid-flipping, take-
off of the entire group is probable, whereas if no other
birds show the action, the sender usually ceases lid-flipping and remains on the surface with the rest of the group.

Signals that coordinate movements of groups may be
expected to show two kinds of indexical qualities. The
first is indication of the *intended direction* of movement,
and such signals may be very simple. For example, Kummer
(1968) reports that Hamadryas baboons move off together
after the night's sleep. The males face and start out in
some direction, each with his harem, in an apparent attempt
to recruit others. After several false starts, eventually
the entire group moves in the direction indexed by a particular male. Of course, this simple signal is also iconic
because it reveals the intention of the male to move as
well as indexing the direction of intended movement.

The other kind of indexical signal integrates the
spatial organization of the moving group. For example,
Keenleyside (1955) cites the black mark on the dorsal fin
of *Pristella riddlei* as promoting schooling in these fish
(*fig 8-10*, next page), and Shaw (*e.g.*, 1970) has investigated schooling responses experimentally with a striped

Fig 8-10. Indexic signals used in cooperative behavior. At left, the invitation for allogrooming behavior of the lesser kudu (after Walther, 1964) and at right the black fin-markings of Pristella riddlei *that promote schooling (after Keenleyside, 1955).*

drum such as used in flicker studies *(ch 5)*.

Fishes such as the alewife, menhaden and gizzard shad have black marks just behind the gill covers , whereas others such as the red drum have a spot on the caudal pe- duncle.   Many large tuna have a conspicuously dark pector- al fin (unusual among fishes, which often have transparent pectoral fins).   If these various markings are all indexic signals that help a fish position itself with respect to companions, the body locus of such signals should be in- vestigated in relation to the geometry of schools.

The markings and geometry of schools may depend upon environmental circumstances.   In reviewing the behavior of cichlids, G.W. Barlow (1974: 19) remarks that "In gen- eral, the more the fish finds itself in open water, often schooling, the greater the tendency to develop the row of spots into a stripe.   While up in the water but over rocks, particularly hovering in groups, the general pattern is for the appearance of the mid-body and base-of-tail spots. When the fish move closer to the bottom, often passing in and out among holes or submerged tree branches, one sees a combination of spots and bars with softly developed edges."

Baylis (1974a) provides photographs of the cichlid fish *Neetroplus nematopus*, which may be gray with a prom- inent mid-body bar that is either black or white.   The black bar is shown in schooling whereas the white bar is shown by parents guarding young, when they repulse all other adults (G.W. Barlow, 1974: 21 and Baylis, *pers. comm.*). This example suggests Darwinian antithesis in social *vs.* nonsocial behavior, a type of antithesis apparently not

---

heretofore recognized.

special group activities

Sometimes animals cooperate toward some collective
end that requires for its success special kinds of coordination among individuals, as in cooperative hunting
by prides of lions or cooperative anti-predator reactions
in flocks of birds. Such specific coordination is more
likely to be peculiar to the species than are the simple
group movements discussed above. There are, of course,
heterospecific foraging groups (*e.g.*, Moynihan, 1962;
Morse, 1970), but these rarely involve a sophisticated degree of cooperation among component individuals. The success of a special group activity may require specific behavioral tactics, as in special aerial tactics in bird
flocks approached by a raptor (*e.g.*, Hailman, 1959a). In
such cases, *species-specificity* of signals may be expected,
an example of which is the wing speculum patterns of pud-
dle ducks (see figure 20 in Hailman, 1977a), which help
assemble homospecific groups in flight during emergency
situations.

When special activities revolve around some specific
place in the environment, such as a predator or food sup-
ply, one may expect the use of indexic signals to point out
the specific places or coordinate special movements with respect to the places. For example, the direction of gaze by
an animal giving an alarm call may index the position of
the source of alarm (see *fig 8-1a*), or the direction of
movement of a fleeing white-tailed deer may become conspicuous
by the white tail, which indexes simultaneously the probable direction of disturbance as well as the intended direction of fleeing. Of course, it is possible that animals
give "false alarms" to deceive their companions and thereby gain some benefit (Charnov and Krebs, 1975).

Some characteristics of signals in cooperative behavior are summarized in *table 8-VII* (next page), which like
preceding tables in this chapter is meant to suggest kinds
of relevant variables, rather than exhaustively review animal signals.

a final note on pragmatics

It would certainly be destructive to my aims if this

Table 8-VII

Optical Characteristics of Cooperative Signals

Behavior	Signal Function	Specific Character-istics
social maintenance activities	invitation	maintenance icons
	address	orientational indexes
	designate site	indexic signals
coordinated group movement	promote movement	locomotory icons
	direct movement	indexic signals
	position individuals within group	indexic signals
special group activities	species-specificity	?
	coordination of individuals	indexic signals

brief discussion on pragmatics and the classification of social behavior used to structure it were taken by any reader as a view of how social behavior and communication are organized in animals. Throughout this discussion, real social problems of behavior have been crushingly simplified in order to extract some main features of optical signals from their broad pragmatic contexts. A huge literature in ethology could be cited to allay any false impressions, but let me cited just three recent papers that might be consulted to appreciate the complexities of social interactions and use of optical signals therein.

Simpson (1973) provides a useful analysis of how recognition of individual animals changes the meaning of displays as well as the contexts in which they are given. Beer (1975) makes a sensitive analysis of signals by gulls, reviewing in some cases how analytical explanations have changed through the studies of G.K. Noble, N. Tinbergen and others to the present day. McKinney (1975) performs a similar service in reviewing the complicated factors that dictate the use of

optical signals in ducks, which also have a long history
of analysis from O. Heinroth, through K. Lorenz and U.
Weidmann and many others. These kinds of studies are
pioneering the way to an understanding of pragmatics in
optical communication. My aim in this section has been
merely to show that optical properties of signals may be
dictated by their pragmatic use, and, conversely by implication, that the study of optical principles in signaling
may help to unravel the complexities of communication e-
vents in social animals.

Overview

The kind of information transferred and the way in
which it is encoded yield many predictions about the characteristics of optical signals. This semiotic approach
to signals begins with the relations of signals to their
referents, where there are three semantic relationships:
indexic relations, where signals literally point out their
referents; iconic relations, where they resemble their referents; and symbolic relations, where they stand for their
referents in any other way held in common by sender and
receiver. All behavioral and morphological elements of
optical signals may be indexic or symbolic, and all except
possibly orientations may be iconic. Most of the highly
ritualized movements (including gestures) and shapes (including postures) are iconic predictors of unritualized
behavior that is likely to be shown subsequently. Certain
optical characteristics of indexic signals are to be expected, but there are no simple generalizations about symbolic signals. The second semiotic problem, the syntactic
relations among signals, is a rich and unexplored area
that may be partitioned into problems of coding, destination and interpretation. Coding problems lead to expectations of signal-economy and Darwinian antithesis in optical
signals; destinational problems to encephalization of signals and the importance of eye-contact; and interpretational problems to particularly interesting areas of "metacommunication," punctuation and modification by signals, including quantification by signal "intensity." Finally, the
pragmatic aspect of semiotics was treated only briefly to
see how consequences or uses of optical signals affect
their characteristics. Some expectations about the characteristics of signals used in cooperative behavior, agonistic

contests, courtship and parental behavior may be extracted, but generalities are few because of the vast differences in behavior among species. Pragmatics provides the most useful general framework for putting together the principles of this volume with the aim of predicting the optimum characteristics of optical signals used for specific purposes in specific environments.

Recommended Reading and Reference

Intraspecific communication is by definition a social phenomenon so that the most instructive sources of information about pragmatics of optical signals may be found in general reviews of social behavior. Two excellent volumes --J.L. Brown's (1975) *Evolution of Behavior* and Wilson's (1975) *Sociobiology*--provide a good entry into the literature. N. Tinbergen's (1951) classic *Study of Instinct* and Darwin's (1872) *Expressions of Emotions* are still worth careful reading. Among textbooks, *Mechanisms of Animal Behavior* (Marler and Hamilton, 1966) provides the most complete coverage of the subject, although much valuable material also occurs in more recent texts. W.J. Smith (1977) has an important overview of communicational behavior due to appear about the time of the present book, and Sebeok's (1977) authoritative compendium is due to appear shortly; both will prove valuable resources for all persons interested in animal communication.

Chapter 9

CONCLUSION

... the other is a conclusion, showing from
various causes why the execution has not
been equal to what the author promised to
himself and to the public. --Johnson

Some concluding words are appropriate to chart the
problems and prospects for the analysis of animal communication, especially via optical signals. What I believe
that I have shown is this. Analysis of communication must
begin with a statement of an epistemological framework
(ch 1) that includes not only general methods of science,
but also the breadth of causal factors required for explanation at a chosen level, *e.g.*, the causes and origins
of behavior in general: control, ontogeny, preservation
and phylogeny. Next, the task becomes one of articulating
some tentative working scheme of communicational phenomena, such as the blend of semiotics, cybernetics and e-
thology proposed in *ch 2*. Analysis may proceed to scrutinize properties of the channel *(ch 3)*, the sender *(ch 4)*
and the receiver *(ch 5)* as basic elements in the communicational process. It is then possible to investigate
the problems of transmitting a signal in the presence of
particular kinds of noise *(ch 7)*, an investigation poten-
tiated by a diversion into specialized signals designed
to deceive certain receivers *(ch 6)*. Finally, the kind
of information being transmitted and the way in which it
is encoded constitute the apical matter of concern *(ch 8)*.
The question I cannot answer is whether this set of factors is sufficient to explain communication, or even the
design of signals used for communication in a particular

channel.

Having pursued this course with regard to the design
of optical signals, I can state that unrecognized issues
were uncovered, hazy phenomena were brought into better
focus, and some well understood problems were organized
into a more coherent framework. The terminal task is
therefore to review some problems that exist and estimate
the prospects for future study of optical signals.

Problems

Although this book is not meant as a review of the
literature on optical signals, the literature reviewed
as background for articulating principles did point up
some recurring problems. These problems are conveniently
discussed within the framework of the classes of causes
and origins of behavior *(table 1-II)*. Elsewhere, I have
provided an overview of all four classes of behavioral
determinants for various aspects of signaling by reflected light (Hailman, 1977a); here the objective is merely
to discuss certain problems about optical signals *per se*
within that same framework.

preservation: species-specificity and mimicry

Preservation of a signal within a population of animals is primarily due to natural selection, kinship selection and related factors (see *Pragmatics* section of
ch 8). By studying control aspects of communication (see
table 1-II, p. 13), one is often able to guess at the a-
daptive significance of signal-characteristics. A few
such commonly employed guesses, however, seem overused,
particularly those relating to species-specificity or its
absence and those relating to mimicry.

There are definite communicative situations in which
one expects to find some species-specificity in optical
signals due to their pragmatic uses *(ch 8)*. Among these
are signals that serve as behavioral barriers to interspecific hybridization (ethological isolating mechanisms)
and signals that promote species-specific group actions
of special kinds. As pointed out in *ch 8*, however, many
pairs of closely related species that run a high risk of
potential interbreeding differ in only minor, although
consistent, ways. I think a more viable research strategy

for explaining differences among species is to begin looking at other differences in the communicational situations:
differences in the pragmatic uses of the signals *(ch 8)*,
differences in the signaling environment *(ch 7)* and so on.
It seems unlikely, for example, that the multiple differences in coloration of male wood warblers are required as
simple redundancy to insure recognition of the species.
Burtt (1977) has made laudable progress in understanding
specific differences in various aspects of coloration in
these birds, although much more detailed study will be required to account for all the differences that occur.

Second, the other side of the species-distinctiveness
coin is the convergence of some signals of different species to highly similar forms. Cody (1969) and Moynihan
(1968) have presented arguments that species often evolve
to a common type in order to promote interspecific communication. Certain woodpeckers, for example, are so similar that each species excludes the other from its territory where the ranges overlap. Although species may in
such cases evolve toward common types for interspecific
communication, this explanation for species-similarities
in signals runs the risk of being overused. In some clear-
cut cases convergence cannot possibly be due to this factor. For example, there occurs in Africa a bird that looks
strikingly like the eastern meadowlark of North America,
yet is in an entirely different family. The nearly identical color patterns of the two species must be due to
convergence for signaling similar things in similar environments, since their ranges do not overlap. Nor does
the similarity stop with optical signals. J.T. Emlen
(pers. comm.) played tape recordings of the North American
bird to the African species and received strong responses:
the birds also sound alike. There are many other examples
of unrelated look-alikes among birds whose ranges do not
overlap. With such striking convergence in allopatric
species why should one assume that convergence in sympatric
species is due to a different cause? One should look
first for similarities in the signaling situation (kinds
of environmental noise, kinds of information being communicated, *etc.*) as an explanation of convergence in signals.

Third, with all due respect to those who have studied
the phenomenon carefully, I believe that mimicry is an
overused explanation of optical signal-characteristics.
As Wickler (1968) has pointed out, especially in his discussion of the "coral-snake" patterns of snakes, mimicry

is a complicated phenomenon that is difficult to establish with certainty. Although I am convinced that mimicry does occur, and indeed occurs widely in some groups for some purposes, other off-hand suggestions of mimicry bear closer scrutiny. For example, eye-spot patterns occur widely as markings in animals. In some cases, as with the marks on underwings of certain moths (Blest, 1957), the markings appear as real mimicry designed to frighten birds by true deception. However, one should note that the eye-spot pattern in general also maximizes many properties of visual conspicuousness *(ch 7)*. Especially where such spots do not occur as a pair (*e.g.*, on the gold-spotted eel), it seems unlikely that mimicry of eyes has played an important role in evolution, unless specific evidence for eye-mimicry can be found. In general, when there is no compelling and reasonably direct evidence for mimicry, one should seek to evaluate a range of possible explanations for a given color pattern.

These three issues have a common thread: all involve explaining the characteristics of optical signals with reference to *other species*. By off-handedly assigning interspecific explanations to communicational signals that have not been analyzed with respect to the species' own needs and environment, one becomes blinded to basic factors that dictate the design of signals. The interspecific explanations should emerge as phenomena of last resort, since the commonest problem that faces each species is reliable communication among its own members in a noisy channel.

phylogeny: icons, deception and mimicry

A person trained in semiotics might assert that iconic optical signals often evolve from incipient movements of the referent behavior because those incipient movements resemble the behavior and therefore are iconic. Ethologists have tended to write as if something like the reverse were true: optical signals are iconic because they evolve from incipient movements of their referents. In this second view, the resemblance is virtually accidental, a happy artifact that allows the ethologist to trace probable phylogenetic origins of optical signals. In fact, a combination of both views is probably necessary to understand the phylogeny of most optical signals.

For simplicity, suppose that learning and other forms
of experience play a minor role in the ontogeny of signaling behavior (but see next section), and that the evolution of communication depends upon natural selection
acting on inter-individual variation in behavioral outputs
of the sender and in responsive outputs of the receiver.
The most parsimonious evolutionary hypotheses must assume
that during the course of phylogeny there is a continuous
functional relationship between outputs of the sender and
responses of the receiver: no saltatory evolutionary "jumps"
are allowed, a restriction necessary in order to keep the
hypotheses consistent with current evolutionary thought
(*e.g.*, Mayr, 1963). With this background, it is possible
to sketch hypotheses about the evolution of iconic signals.

In a simple case, incipient movements of locomotion
(*e.g.*, Daanje, 1950) resemble actual locomotion and often
precede it temporally. Therefore, any receiver that would
benefit from being able to predict locomotion by its companions could do so by cuing on the incipient movements,
even though they have not been evolutionarily influenced
by factors relating to their role as semiotic signs *(ch 2);*
also see the discussion by Johnston (1976: 58). Natural
selection then favors individuals that can react as receivers to the incipient movements. It is likely that
they can so react because the incipient movements resemble
full locomotion: they are *already* icons. If the sender
benefits from transferring the predictive information a-
bout his impending locomotion, then individuals that transfer this information most efficiently will be selected for.
In the simplest sense, "most efficiently" means that the
receiver notices the movements and responds in a way ben-
efiting the sender (*e.g.*, coordinate group movement).
This is the classical process of ritualization, in which
the incipient movements are enhanced by factors such as
exaggerated amplitude of movement, increased frequency,
stereotypy, conspicuous coloration and so on.

Second, the evolution of morphological icons proceeds similarly. For example, *ch 8* recounted a whole
series of hypotheses concerning factors that promote the
use of antlers and horns as optical signals in ungulates.
The original evolution of these structures was presumably
guided by their use in defense against predators or fighting among conspecifics. Because they are displayed prominently in preparation for fighting, they are *automatically*
signs without further evolution to enhance their signal

value. Then, receivers that cue on these structures can
better predict impending attack, and ritualization via
increasing the conspicuousness of the structures proceeds
as in the case of behavioral signals, above. In this case,
however, ritualization may include additional morphological
elements that resemble horns and therefore create iconic
redundancy: manes, ear-shape, ear-markings, facial markings and so on (see figure 22 in Hailman, 1977a). Such
additional elements are said to be "automimics" of the
basic structures (Guthrie and Petocz, 1970).

The evolution of mimetic patterns in such intraspecific
communication proceeds because the mimetic patterns are
already iconic signs. For example, the pinnae of the
pronghorn are erected to receive acoustic information in
social interactions, and hence already have a general
shape and orientation that resembles the animal's horns.
Selection favors variant shapes of pinnae that more
strongly resemble horns (*e.g.*, curled "hook" at the tip)
because such shapes are more perfect icons and help increase the redundancy of the optical display. There is
no need to believe that receivers actually mistake the
pinnae for additional horns.

Not all cases of automimicry need involve redundancy,
however. In primate species having optical similarities
between the markings and structures of the anogenital
region and facial region, the duplication cannot be due
to simultaneous redundancy because both sets of signals
usually cannot be seen by a receiver at the same time.
The presumed sequence of phylogeny in these cases begins
with the use of anogenital display as iconic predictors
of behavior, which are then made more conspicuous by coloration. With increasing encephalization of signals for
the many reasons discussed in previous chapters, elements
of the anogenital signals will be duplicated on the facial
region because receivers are *already* cuing on such elements. There is mimicry, but not deception.

Not all cases of reputed mimicry within a species are
necessarily mimicry, either. For example, the markings
on anal fins of certain mouth-brooding cichlids (see *fig
8-2d*, p. 253) have been cited as egg-mimics. Wickler (1968:
226) states that "the female repeatedly attempts to take
the round finspots into her mouth as if they were eggs.
Since the spots are located next to the male sexual aper-
ture, spermatozoa automatically end up in the female's
mouth." Although it is possible that this interpretation

is correct, there are reasons to doubt it, and there are
alternative explanations for the characteristics of the
finspots and their phylogeny. Fish have keen vision and
it seems unlikely that the female *Haplochromis burtoni*
cannot discriminate her own eggs from two-dimensional
markings on her mate's fins. Furthermore, the female her-
self possesses similar markings on her fins, although
they are not as conspicuous as the male's markings. Other
cichlids also possess such markings, but some do not look
like eggs at all (to me, anyway), and it seems likely that
the species-specificity of these markings is selected for
as a reproductive isolating mechanism of last resort: a
final species-identification before the sperm are picked
up and the eggs fertilized in the female's mouth.

Suppose the female *H. burtoni* is not actually attempting to pick up markings that she cannot discriminate
from eggs. How could the evolution of this signal-system
proceed without recourse to explanations involving visual
deception? One possibility is as follows. Nipping at the
fins of the mate is observed frequently in both sexes of
cichlids, and presumably helps stimulate the partner for
spawning. Both sexes may therefore evolve markings that
index the optimum sites of such tactile stimulation: the
area near the genital pore. There are many hypotheses
that would account for the circularity of these indexic
signals in species in which they are circular: predisposition of receiver-systems to be organized with bullseye-
like receptive fields *(ch 5)*, conspicuousness of regular
geometric patterns *(ch 7)*, conspicuousness due to outlining, internal contrast of color and brightness, *etc.*
Because the male moves forward while releasing sperm, his
anal fin also indexes the position of sperm for the female, and any markings on the fin may enhance the conspicuousness of the index. This explanation is not in-
consistent with Wickler's interpretation to this point:
it is merely a more complete statement of possible steps
in phylogeny.

Wickler *(loc. cit.)* notes that the female of many
species of cichlids pushes the male aside with her mouth
to gather his sperm, and he notes that the "egg-dummies
may be just an insurance that pushing will be converted
to sucking." Therefore, the crux of the question is
whether there is *further* change in the signal to make it
more egg-like in order to deceive the female into making
a different response. Possibly in *H. burtoni* there has
been, but in cichlids whose indexic markings are not egg-

like the female takes the sperm anyway, so it is not logically necessary to postulate true deception. If there is
no true deception, then it is questionable whether mimicry
occurs at all. If one of the attributes of this indexic
signal is species-specificity to prevent hybridization,
then selection pressures acting on various species to be
simply different from all others is likely to produce
egg-like markings in at least one species of the group.
Egg-mimicry might be involved, but the entire signal-system could have evolved with no deception and no mimicry
at all.

Finally, the phylogeny of *inter*specific mimicry is
different yet. Take for example the evolution of eye-
like spots on the underwings of certain moths (Blest,
1957). In this case, the behavior of the moth when confronted by a predator was probably initially dependent
upon simple startle *(ch 6)*, mediated by incipient flight
movements of the wings. The effectiveness of the startle-
movement was then enhanced by conspicuous coloration of
the usually hidden underwing. Any coloration that increased conspicuousness would be favored, and among those
patterns that maximize conspicuousness are bullseye-like
spots. Finally, those conspicuous spots that were actually
mistaken by predators for eyes were further favored because of the additional frightening effect.

The conclusion of this analysis is simply that not
all forms of iconic signals and reputed mimicry evolve in
the same way. Interspecific mimicry begins with selection
for conspicuousness or other attributes, and proceeds toward convergence with another stimulus when variation
allows and the results of the resemblance are favorable
to the sender: mimicry is a phylogenetic endpoint selected
for true deception. Intraspecific resemblance, on the
other hand, may evolve without any deception. They may
arise to create redundancy (weapon mimicry), to duplicate
signals for use elsewhere on the body (anogenital-facial
mimicry) or to enhance conspicuousness by various means.
In intraspecific resemblances, the receiver can usually
distinguish the iconic signal from the thing it resembles,
and in some cases the resemblance may even be accidental.
Entire books on mimicry have been written as if all mimicry and the evolutionary processes giving rise to it were
essentially identical: that is unlikely.

ontogeny: the role of experience

The role of experience in the development of communicational behavior is understudied, and probably for at least two reasons. First, it is convenient in evolutionary thinking to treat behavioral outputs of animals as if they were fixed by the genetic endowment and showed little variation among individuals due to differences other than genetic differences. Indeed, in order to argue the points in the previous section, I explicitly made such an assumption for purposes of simplifying the evolutionary explanations. Second, the ontogeny of behavior is understudied simply because it is difficult to study--not only technically difficult, requiring special rearing conditions and long-term scrutiny, but also because it is conceptually difficult and experimentally difficult to dissociate all the interacting variables.

It seems likely that the form of many of the species-common signals of the kind discussed in this book develop without an appreciable role played by experience. It is not so obvious that the recognition or use of those signals is largely experientially independent. For example, Sackett (1966) found that rhesus macaques reared in isolation from birth responded differentially to photographic transparencies showing different facial expressions of other individuals. This result does not mean, however, that the information transferred by such signals in wild monkey troops is independent of experience. Stephenson (1973) provides five examples of troop-specific use of signals in Japanese macaques. It seems unlikely that these intraspecific differences are due to genetic differences among the troops. More likely, different signals have come to have different referents through individual experience, and these semiotic relationships have persisted in the troops through cultural transmission. If troop-specific communication can be culturally transmitted, species-specific communication can be also: it is only necessary that all individuals have roughly the same experiences during ontogeny.

Experiential factors in communication are by no means restricted to primates, where they are not unexpected. The newly hatched laughing gull chick responds appropriately to the parental signal for feeding: the red bill-tip held in a specific orientation, moved in a specific direction and at a specific speed (see figure 15 in Hailman,

1977a). However, chicks in the wild also respond to many
*in*appropriate stimuli, and then as a function of time
since hatching decrease their responsiveness to inappro-
priate objects (Hailman, 1967a). Furthermore, experiments
demonstrate that chicks develop a much fuller perceptual
ideal: newly hatched chicks respond to simple models of
the parent's bill whereas experienced chicks demand a
life-like model of the parent (see figure 18 in Hailman,
1977a). Experiments show that food-reinforcement can promote relevant kinds of perceptual learning in these gull
chicks (Hailman, 1967a, 1971). In short, the physical
object that elicits responsiveness (*i.e.*, the parent) does
not change, but the key sign-vehicle to which chicks respond does change rather markedly during ontogeny, a process I called "perceptual sharpening." Indeed, what Sackett's (1966) study with rhesus macaques may really show
is that young monkeys have an initial disposition to respond differently to different facial expressions; perhaps those first responses are then ordinarily channeled
ontogenetically by experiences in real social communication with other individuals.

Given that experience can play a role in optical
communication, one must still ask how this fact is relevant to the design-characteristics of optical signals.
Although this problem-area has been little studied, there
are at least two kinds of potential answers. First, the
behavioral and morphological elements of the parent gull's
signal are designed primarily to elicit the first responses
from newly hatched chicks. If the early responses can be
assured, the chicks learn more sophisticated sign-vehicles.
What they learn may be entirely environmentally dependent,
with one configuration being as readily learned as the
next. Using food as reward for pecking, for example, I
conditioned one species of gull chick to prefer models of
a different species (Hailman, 1967a). The physical signal is structured primarily to insure the early responses
that initiate the ontogenetic sequences.

A second possibility was suggested years ago (Hailman, 1959b), although it is pure speculation. If signals
are learned, at least in part, then their characteristics
may be structured to promote the learning. It is not clear
yet what optical characteristics one would expect of signals that promote visual learning. I suggested that the
crepuscular, hole-nesting wood duck--whose life in the
nest is spent in dim illumination and who communicates as

an adult primarily in twilight--might be particularly dis-
advantaged in learning optical signals of the species.
These factors might therefore play a role in the evolution of the strikingly colorful and complex display plumage of the male. The example might be wrong, but the
possibility remains that certain signal-characteristics
could be easier to learn than others, and hence selected
for to promote perceptual learning.

Prospects

It seems to me that there are three issues of particular importance to consider in conclusion. Now that a
framework of factors potentially important in the design
of optical signals has been laid out, how should one proceed with future studies? Second, what are the implications of signal-characteristics for communicational studies
in general? And last, of what direct relevance is the
study of animal signals to man himself?

strategy for future analysis

It appears unlikely that one could predict the signal-
characteristics of a species from even a detailed knowledge
of the factors shown to be of importance in this volume.
One may hold such prediction as an ultimate goal, but there
is no way to know if it is an attainable goal unless and
until it is attained. In order to make progress, then,
some lesser goal is needed that provides a clear method
for analysis. I believe that the goal is to be able to
predict *differences* in signal-characteristics among species
or other populations. The strategy is to choose groups
for comparison that are as similar as possible in all ways
except for the signal-characteristic under investigation.
For example, if one wants to understand why the male cardinal is so conspicuously red, it seems only sensible to
compare it in detail with the closely related pyrrhuloxia,
whose male has only a wash of red on the breast and whose
female is extremely similar to the female cardinal. One
must ask what kinds of differences exist in the communicational needs and communicational environments of the two
species. There is, it seems to me, some real prospect of
answering this kind of question about species-differences,

whereas asking simply why an animal has the signals it does may often prove frustrating.

The analysis is more complicated than simply comparing animals, of course. One must document how communication takes place (control) and how it develops in the individual (ontogeny) before sense can be made of comparative correlates between signals and other variables. The comparative method then focuses on factors that correlate with the ontogenetic systems that lead to the control endpoints, and these factors may be simple variables of the physical signaling environment or complex factors of the cultural environment. Or, the strong correlation may be with taxonomic status, in which case one suspects that historical factors in phylogeny are responsible for the species-differences. Therefore, even though one is using a comparative method, the ultimate analysis involves all four of the classes of behavioral determinants *(table 1-II)*.

In charting the variables for comparison of optical signals between species, one must take into account all the kinds of problems reviewed in this volume: ambient light falling upon the animal, properties and limitations of the sender's abilities to reflect that light, kinds of information being encoded in the signals, optical noise in the communicating environment, and abilities of the receiver to accept and decode the signals. Furthermore, non-signal hypotheses must be considered as explanations for animal coloration and behavior (see *chs 4* and *6*). Making a comparison between two species is no simple task, but it is one likely to produce new ideas about the optical characteristics of animal signals.

signals and communication

In *ch 8* I emphasized that an understanding of *what* was being communicated could help to understand *how* it was communicated. A major implication of this book, however, reverses that sequence and hence emphasizes the interrelationships between signals and the information they encode. It seems to me a reasonable assumption that if one begins looking in detail at the characteristics of optical signals, those characteristics will help clarify the kind of information being transferred, and, indeed, the entire communicational process. For example, the study by Thomas Stillwell and me *(in prep.)* of inciting by female mallards reveals that it is a clear indexic

---

signal, and not vaguely symbolic as previously considered.
Although we have not studied the responses of the receivers
in the detail required to understand the communication
thoroughly, it is clear that merely the form of the signal itself suggests hypotheses about its informational content. If one is to understand communication, I believe
one needs to ask after the details of the signals themselves.

man's optical signals

Belonging to *Homo sapiens* the author and reader a-
like have some measure of curiosity about this species'
optical signals. Since man is an animal, albeit a complex
one, he should not be totally free of the factors that dictate the design of signals in other animals. He has clearly elaborated these, and in some ways has freed his signals from the kinds of evolutionary constraints affecting
most other species. There have been numerous attempts to
explain some of the basic "body signals" of man that are
similar to the intrinsic signals of other animals. Such
studies by ethologists in particular have produced many
interesting, if not bizarre, hypotheses. This is not the
place to review man's signals and hypotheses about them,
but it is the place to point out that none of these studies
of man has attempted to analyze his signals from the optical viewpoint taken in this volume. Such a viewpoint holds
promise for better understanding of facial expressions,
gestures and other intrinsic optical signals.

Man, however, has created a whole array of extrinsic
optical signals. These range from very straightforward
signals such as traffic lights, to complex creations such
as dance, architecture, painting and even the notational
systems for written language and mathematics. This great
range of extrinsic signals cannot be understood in most
cases without considerable study of culture and linguistics,
but it should be possible to begin examining extrinsic
signals from the optical viewpoint.

Epilogue

Samuel Johnson was correct. I had hoped to find some
magic key that would unlock the major secrets of communication, and to that I was not equal. The promise to the
public, however, was a more modest goal of articulating
factors important in the design of optical signals; the
success or failure of that attempt is now before the reader.

Suggested Reading from Scientific American

A consistent comment received by Peter Klopfer and me concerning
our textbook (Klopfer and Hailman, 1967) pertained to the usefulness of
the appendix that provided a list of articles from the monthly magazine
Scientific American. For the convenience of the reader who is not technically involved with subjects treated in this volume, I have compiled
a selected list of articles, grouped by chapter. A few articles are included under two or more chapters. I have avoided many interesting but
tangential articles (e.g., a large number on lasers), selected from an
array in other areas (e.g., human visual perception) and cited only the
most recent articles on still other topics (e.g., mechanisms of vision).
In many cases, there has been no article in the last decade that treats
certain material mentioned in a chapter, and there are no central articles for chs 7 and 9. The articles listed below are not included in
the terminal bibliography of the book unless specifically cited in the
text.

1. INTRODUCTION

The evolution of behavior, K.Z. Lorenz, Dec 1958.
The evolution of behavior in gulls, N. Tinbergen, Dec 1960.
Mathematics in the biological sciences, E.F. Moore, Sep 1964.
How an instinct is learned, J.P. Hailman, Dec 1969.
Prematurity and uniqueness in scientific discovery, G.S. Stent, Dec 1972.
Confirmation, W.C. Salmon, May 1973.

2. COMMUNICATION

Error-correcting codes, W.W. Peterson, Feb 1962.
Dialects in the language of the bees, K. von Frisch, Aug 1962.
Pheromones, E.O. Wilson, May 1963.
Redundancy in computers, W.H. Pierce, Feb 1964.
Information, J. McCarthy, Sep 1966.
The evolution of bee language, H. Esch, Apr 1967.
Energy and information, M. Tribus and E.C. McIrvine, Sep 1971.
Communication, J.R. Pierce, Sep 1972.
Animal Communication, E.O. Wilson, Sep 1972.

3. THE CHANNEL

The Michelson-Morley experiment, R.S. Shankland, Nov 1964.
Light-emitting semiconductors, E.F. Morehead, Jr., May 1967.
The optical properties of materials, A. Javan, Sep 1967.
Light, G. Feinberg, Sep 1968.
How light interacts with matter, V.F. Weisskopf, Sep 1968.
How light is analyzed, P. Connes, Sep 1968.
Monomolecular layers and light, K.H. Drexhage, Mar 1970.

Optical interference coatings, P. Baumeister and G. Pincus, Dec 1970.
The flow of energy in the biosphere, D.M. Gates, Sep 1971.
Mirages, A.B. Fraser and W.H. Mach, Jan 1976.
The mass of the photon, A.S. Goldhaber and M.M. Nieto, May 1976.

4. THE SENDER

Carotenoids, S. Frank, Jan 1956.
Hormones and skin color, A.B. Lerner, Jul 1961.
Biological luminescence, W.D. McElroy and H.H. Seliger, Dec 1962.
The evolution of bowerbirds, E.T. Gilliard, Aug 1963.
The flight muscles of insects, D.S. Smith, Jun 1965.
The swimming energetics of salmon, J.R. Brett, Aug 1965.
The mechanism of muscular contraction, H.E. Huxley, Dec 1965.
Inhibition in the central nervous system, V.J. Wilson, May 1966.
The energetics of bird flight, V.A. Tucker, May 1969.
How is muscle turned on and off?, G. Hoyle, Apr 1970.
Reflectors in fishes, E. Denton, Jan 1971.
The sources of muscular energy, R. Margaria, Mar 1972.
How we control the contraction of our muscles, P.A. Merton, May 1972.
The flying leap of the flea, M. Rothschild *et al.*, Nov 1973.
The effects of light on the human body, R.J. Wurtman, Jul 1975.
Synchronous fireflies, J. and E. Buck, May 1976.
Flashlight fishes, J.E. McCosker, Mar 1977.

5. THE RECEIVER

Experiments in discrimination, N. Guttman and H.I. Kalish, Jan 1958.
How we see straight lines, J.R. Platt, Jun 1960.
Shadows and depth perception, E.H. Hess, Mar 1961.
Experiments in animal psychophysics, D.S. Blough, Jul 1961.
Aftereffects in perception, W.C.H. Prentice, Jan 1962.
Experiments with goggles, I. Kohler, May 1962.
The moon illusion, L. Kaufman and I. Rock, Jul 1962.
Visual pigments in man, W.A.H. Rushton, Nov 1962.
The perception of neutral colors, H. Wallach, Jan 1963.
Inhibition in visual systems, D. Kennedy, Jul 1963.
Afterimages, G.S. Brindley, Oct 1963.
The visual cortex of the brain, D.H. Hubel, Nov 1963.
Vision in frogs, W.R.A. Muntz, Mar 1964.
The illusion of movement, P.A. Kolers, Oct 1964.
Three-pigment color vision, E.F. MacNichol, Jr., Dec 1964.
Texture and visual perception, B. Julez, Feb 1965.
Night blindness, J.E. Dowling, Oct 1966.
Vision and touch, I. Rock and C.S. Harris, May 1967.
Molecular isomers in vision, R. Hubbard and A. Kropf, Jun 1967.
How light interacts with living matter, S.B. Hendricks, Sep 1968.
The processes of vision, U. Neisser, Sep 1968.
Visual illusions, R.L. Gregory, Nov 1968.
The processing of visual images, C.R. Michael, May 1969.
How we remember what we see, R.N. Haber, May 1970.
Visual cells, R.W. Young, Oct 1970.
Advances in pattern recognition, R.G. Casey and G. Nagy, Apr 1971.

Eye movements and visual perception, D. Norton and L. Stark, Jun 1971.
Multistability in perception, F. Attneave, Dec 1971.
Nonvisual light reception, M. Menaker, Mar 1972.
The neurophysiology of binocular vision, J.D. Pettigrew, Aug 1972.
The control of sensitivity in the retina, F.S. Werblin, Jan 1973.
The infrared receptors of snakes, R.I. Gamow and J.F. Harris, May 1973.
The perception of disoriented figures, I. Rock, Jan 1974.
The perception of transparency, F. Metelli, Apr 1974.
Contrast and spatial frequency, F.W. Campbell and L. Maffei, Nov 1974.
Visual pigments and color blindness, W.A.H. Rushton, Mar 1975.
Experiments in the visual perception of texture, B. Julez, Apr 1975.
Visual motion perception, G. Johansson, Jun 1975.
The perception of surface color, J. Beck, Aug 1975.
The resources of binocular perception, J. Ross, Mar 1976.
Subjective contours, G. Kanizsa, Apr 1976.
Polarized-light navigation by insects, R. Wehner, Jul 1976.
Visual cells in the pons of the brain, A.R. Gibson, Nov 1976.
Negative aftereffects in visual perception, O.E. Favreau and M.C. Cor-
   ballis, Dec 1976.

6. DECEPTION

Cleaning symbiosis, C. Limbaugh, Aug 1961.
Orchids, J. Arditti, Jan 1966.
Reflectors in fishes, E. Denton, Jan 1971.
Communication between ants and their guests, B. Hölldobler, Mar 1971.
Mimicry in parasitic birds, J. Nicolai, Oct 1974.
Moths, melanism and clean air, J.A. Bishop and L.M. Cook, Jan 1975.
The causes of biological diversity, B. Clarke, Aug 1975.

8. INFORMATION

The evolution of behavior, K.Z. Lorenz, Dec 1958.
The evolution of behavior in gulls, N. Tinbergen, Dec 1960.
The social life of baboons, S.L. Washburn and I. DeVore, Jun 1961.
The fighting behavior of animals, I. Eibl-Eibesfeldt, Dec 1961.
Chimpanzees in the wild, A. Kortlandt, May 1962.
The schooling of fishes, E. Shaw, Jun 1962.
The evolution of bowerbirds, E.T. Gilliard, Aug 1963.
The origins of facial expressions, R.J. Andrew, Oct 1965.
The evolution of bee language, H. Esch, Apr 1967.
Visual isolation in gulls, N.G. Smith, Oct 1967.
The phalarope, E.O. Höhn, Jun 1969.
How an instinct is learned, J.P. Hailman, Dec 1969.
The social order of turkeys, C.R. Watts and A.W. Stokes, Jun 1971.
Animal communication, E.O. Wilson, Sep 1972.
The migrations of the shad, W.C. Leggett, Mar 1973.
The social system of lions, B.C.R. Bertram, May 1975.
Stomatopods, P.L. Caldwell and H. Dingle, Jan 1976.
Synchronous fireflies, J.and E. Buck, May 1976.
The social order of Japanese macaques, G.G. Eaton, Oct 1976.

All literature cited in the text, including *Recommended Reading and Reference*, is provided here, but articles cited in the *APPENDIX* are not. Numbers in parentheses following each reference are the pages on which the book or article is cited.

Adler, K.K. 1970. The role of extraoptic photoreceptors in amphibian rhythms in salamanders. *Science, 164:* 1290-1292. *(120)*

Albrecht, H. 1962. Die Mitschattierung. *Experientia, 18:* 284-286. *(196)*

Alexander, R. McN. 1967. *Functional Design in Fishes.* Hutchinson Univ. Press, London. 160 pp. *(186)*

Andrew, R.J. 1956. Intention movements of flight in certain passerines, and their use in systematics. *Behaviour, 10:* 179-204. *(295)*

Armstrong, E.A. 1947. *Bird Display and Behaviour*, 2nd ed. Drummond, London. 431 pp. *(292)*

Armstrong, E.A. 1964. Distraction display. *In:* A.L. Thompson (ed), *A New Dictionary of Birds*, McGraw-Hill, N.Y., p. 205. *(178, 180)*

Ashby, W.R. 1961. *An Introduction to Cybernetics.* Chapman & Hall, London. 295 pp. *(34, 43, 53)*

Attneave, F. 1954. Informational aspects of visual perception. *Psychol. Rev., 61:* 183-193. *(261)*

Attneave, F. 1959. *Applications of Information Theory to Psychology.* Rinehart & Winston, N.Y. 98 pp. *(32-33, 53)*

Averill, C.K. 1923. Black wing tips. *Condor, 25:* 57-59. *(112)*

Baerends, G.P. and J.M. Baerends-van Roon. 1950. An introduction to the study of the ethology of cichlid fishes. *Behav. Suppl., 1:* 1-243. *(275, 276)*

Barlow, G.W. 1967. The functional significance of the split-head color pattern as exemplified in a leaf fish, *Polycentrus schomburgkii. Ichthyol. Aquar. J., 39*(2): 57-70. *(166)*

Barlow, G.W. 1974. Contrasts in social behavior between Central American cichlid fishes and coral-reef surgeon fishes. *Amer. Zool., 14:* 9-34. *(171, 196, 231, 296)*

Barlow, G.W. 1977. Modal action patterns. Chapter 7 *in:* T.A. Sebeok (ed), *How Animals Communicate*, Indiana Univ. Press, Bloomington, *in press.* *(33, 263)*

Barlow, G.W. and R.F. Green. 1970. Effect of relative size of mate on color patterns in a mouthbreeding cichlid fish, *Tilapia melanotheron. Comm. Behav. Biol., 4:* 71-78. *(268)*

Barlow, H.B. 1961a. The coding of sensory messages. *In:* W.H. Thorpe and O.L. Zangwill (eds), *Current Problems in Animal Behaviour*, Cambridge Univ. Press, pp. 331-360. *(261)*

Barlow, H.B. 1961b. Possible principles underlying the transformations of sensory messages. *In:* W.A. Rosenblith (ed), *Sensory Communication*, M.I.T. Press, Cambridge, Mass. *(261)*

Barlow, H.B., R.M. Hill and W.R. Levick. 1964. Retinal ganglion cells responding selectively to direction and speed of image motion in the rabbit. *J. Physiol., 173:* 377-407. *(148-149)*

Bartley, S.H. 1951. The psychophysiology of vision. *In:* S.S. Stevens (ed), *Handbook of Experimental Psychology*, John Wiley & Sons, N.Y., pp. 921-984. *(141, 142)*

Bateson, G. 1956. The message "this is play." *In:* B. Schaffner (ed),

Group Processes, J. Macy Foundation, N.Y. (*266, 267*)

Batschelet, E. 1975. *Introduction to Mathematics for Life Scientists*, 2nd ed. Springer-Verlag, Berlin. 643 pp. (*16,17*)

Baylis, J.R. 1974a. The behavior and ecology of *Herotilapia multispinosa* (Teleostei, Cichlidae). *Z. Tierpsychol.*, *34*: 115-146. (*196, 231, 296*)

Baylis, J.R. 1974b. *Neetroplus nematopus*. Amer. Cichlid Associ., 2 pp. (*264*)

Baylis, J.R. 1976. A quantitative study of long-term courtship: I. Ethological isolation between sympatric populations of the midas cichlid, *Cichlasoma citrinellum*, and the arrow cichlid, *C. zaliosum*. *Behaviour*, *59*: 59-69. (*46*)

Beer, C.G. 1961. Incubation and nest building behaviour of black-headed gulls. I. Incubation behaviour in the incubation period. *Behaviour*, *18*: 62-106. (*39, 292*)

Beer, C.G. 1962. Incubation and nest building behaviour of black-headed gulls. II. Incubation behaviour in the laying period. *Behaviour*, *19*: 283-304. (*39*)

Beer, C.G. 1975. Multiple functions and gull displays. *In:* G. Baerends, C. Beer and A. Manning (eds), *Function and Evolution in Behaviour*, Oxford Univ. Press, pp. 16-54. (*289, 290, 298*)

Bekoff, M. 1975. The communication of play intention: are play signals functional? *Semiotica*, *15*: 231-240. (*266*)

Bekoff, M. 1976. Animal play: problems and perspectives. Chapter 4 *in:* P.P.G. Bateson and P.H. Klopfer (eds), *Perspectives in Ethology*, vol. 2, pp. 165-188, Plenum, N.Y. (*268*)

Berry, K. 1974. The ecology and social behavior of the chuckwalla, *Sauromalus obesus* Baird. *Univ. Calif. Publ. Zool.*, *101*: 1-60. (*288*)

Bishop, J.A. and L.M. Cook. 1975. Moths, melanism and clean air. *Sci. Amer.*, *232*(1): 90-99. (*171*)

Blest, A.D. 1957. The function of eyespot patterns in the Lepidoptera. *Behaviour*, *11*: 206-256. (*176, 306, 310*)

Blough, D.S. 1957. Spectral sensitivity in the pigeon. *J. Opt. Soc. Amer.*, *47*: 827-833. (*119, 126, 129*)

Böhlke, J.E. and C.C.G. Chaplin. 1968. *Fishes of the Bahamas and Adjacent Tropical Waters*. Philadelphia Acad. Nat. Sci. 771 pp. (*162*)

Bridgman, P.W. 1927. *The Logic of Modern Physics*. MacMillan, N.Y. (*6, 18*)

Bridgman, P.W. 1938. *The Intelligent Individual and Society*. MacMillan, N.Y. (*18*)

Brown, J.E. and J.A. Rojas. 1965. Rat retinal ganglion cells: receptive field organization and maintained activity. *J. Neurophysiol.*, *28*: 1073-1090. (*149*)

Brown, J.L. 1963. Ecogeographic variation and introgression in an avian visual signal: the crest of the Steller's jay, *Cyanocitta stelleri*. *Evolution*, *17*: 23-39. (*219*)

Brown, J.L. 1964. The integration of agonistic behavior in the Steller's jay *Cyanocitta stelleri* (Gmelin). *Univ. Calif. Publ. Zool.*, *60*: 223-328. (*268, 277*)

Brown, J.L. 1975. *The Evolution of Behavior*. W.W. Norton, N.Y. (*12, 53, 268, 270-271, 273, 300*)

Brush, A.H. 1972. Review of: J. Dyke (1971). *Auk*, *89*: 679-681. (*100*)

Brush, A.H. 1970. Pigments in hybrid, variant and melanic tanagers (birds). *Comp. Biochem. Physiol.*, *36*: 785-793. (*100, 110*)

Burtt, E.H., Jr. 1977. The coloration of wood warblers (Parulidae). Ph.D. Thesis, Univ. of Wisconsin-Madison. (*79, 110, 111, 112, 114,*

139, 171, 217, 228, 242, 305)

Campbell, F.W. and L. Maffei. 1974. Contrast and spatial frequency. *Sci. Amer.*, *231*(5): 106-111, 113-114. *(144)*

Carpenter, C.R. 1964. *Naturalistic Behavior of Nonhuman Primates.* Penn. State Univ. Press, University Park. *(266)*

Catania, A.C. 1964. On the visual acuity of the pigeon. *J. Exp. Anal. Behav.*, *7:* 361-366. *(143)*

Chance, M.R.A. 1962. An interpretation of some agonistic postures: the role of "cut-off" acts and postures. *Symp. Zool. Soc. London*, *8:* 71-89. *(265)*

Charnov, E.L. and J.R. Krebs. 1975. The evolution of alarm calls: altruism or manipulation? *Amer. Nat.*, *109:* 107-112. *(297)*

Chaplin, C.C.G. 1972. *Fishwatchers Guide to West Atlantic Coral Reefs.* Livingston, Wynnewood, Pa. 65 pp. *(162, 186, 244)*

Cherry, C. 1957. *On Human Communication.* John Wiley & Sons. *(48, 49, 53, 249)*

Chomsky, N. 1957. *Syntactic Structures.* Mouton & Co. *(53)*

Clark, E. 1959. Functional hermaphroditism and self-fertilization in a serranid fish. *Science*, *129:* 215-216. *(290)*

Clarke, B. 1962. Balanced polymorphism and the diversity of sympatric species. *Systemat. Assoc. Publ.*, *4:* 47-70. *(184)*

Clarke, T.A. 1970. Territorial behavior and population dynamics of a pomacentrid fish, the garibaldi, *Hypsypops rubicunda*. *Ecol. Monogr.*, *40:* 189-212. *(259)*

Clayton, R.K. 1970. *Light and Living Matter, Vol. I: The Physical Part.* McGraw-Hill, N.Y. 148 pp. *(88)*

Cody, M.L. 1969. Convergent characteristics in sympatric populations: a possible relation to the interspecific territoriality. *Condor*, *71:* 222-239. *(274, 305)*

Conant, R. 1975. *A Field Guide to Reptiles and Amphibians of Eastern and Central North America*, 2nd ed. Houghton Mifflin, Boston. 429 pp. *(178)*

Coren, S. 1972. Subjective contours and apparent depth. *Psychol. Rev.*, *79:* 359-367. *(145)*

Cott, H.B. 1957. *Adaptive Coloration in Animals.* Methuen, London. *(157, 161, 165, 166, 169, 189)*

Coulson, K.L. 1975. *Solar and Terrestrial Radiation: Methods and Measurements.* Academic Press, N.Y. 336 pp. *(88)*

Crews, D. 1975. Effect of different components of male courtship behaviour on environmentally-induced ovarian recrudescence and mating preferences in the lizard *Anolis carolinensis*. *Anim. Behav.*, *23:* 349-356. *(27)*

Crozier, W.J. 1940. On the law for minimal discrimination of intensities. IV. ΔI as a function of intensity. *Proc. Natl. Acad. Sci.* (U.S.), *26:* 382-389. *(122)*

Cullen, E. 1957. Adaptations in the kittiwake to cliff-nesting. *Ibis*, *99:* 275-302. *(14-15)*

Daanje, A. 1950. On locomotory movements in birds and the intention movements derived from them. *Behaviour*, *3:* 49-98. *(295, 307)*

Dane, B. and W.G. Van der Kloot. 1964. An analysis of the display of the goldeneye duck (*Bucephala clangula* (L.)). *Behaviour*, *22:* 282-328. *(263)*

Dane, B., C. Walcott and W.H. Drury. 1959. The form and duration of display actions of the goldeneye (*Bucephala clangula*). *Behaviour*, 14: 265-281. (*263*)

Dartnall, H.J.A. 1953. Interpretation of spectral sensitivity curves. *Brit. Med. Bull.*, 9: 24-30. (*118*)

Dartnall, H.J.A. 1957. *The Visual Pigments*. Methuen, London. 216 pp. (*154*)

Dartnall, H.J.A. (ed). 1972. *Handbook of Sensory Physiology*, Vol. VII/1. *Photochemistry of Vision*. Springer, Berlin. (*154*)

Darwin, C. 1871. *The Descent of Man, and Selection in Relation to Sex*. Appleton, N.Y., Vol. I, 409 pp.; Vol. 2, 436 pp. (*275, 283*)

Darwin, C. 1872. *Expression of the Emotions in Man and Animals*. Appleton, N.Y. (*106, 264, 300*)

Daumer, K. 1956. Reizmetrische Untersuchung des Farbensehens der Bienen. *Z. vergleich. Physiol.*, 38: 413-478. (*195-196*)

Dawkins, R. 1976. Hierarchical organisation: a candidate principle for ethology. Chapter 1 *in*: P.P.G. Bateson and R.A. Hinde (eds), *Growing Points in Ethology*, Cambridge Univ. Press, pp. 7-54. (*261*)

Denton, E.J. and J.A.C. Nichol. 1965. Studies of the reflexion of light from silvery surfaces of fishes, with special reference to the bleak, *Alburnus alburnus*. *J. Mar. Biol. Assoc. U.K.*, 45: 683-703. (*115*)

Denton, E.J. and J.A.C. Nichol. 1966. A survey of reflectivity in silvery teleosts. *J. Mar. Biol. Assoc. U.K.*, 46: 685-722.

Dice, L.R. 1947. Effectiveness of selection by owls of deermice (*Peromyscus maniculatus*) which contrast with their background. *Contr. Lab. Vert. Biol. Univ. Michigan*, no. 34. (*170*)

Dilger, W.C. and P.A. Johnsgard. 1959. Comments on "species recognition" with special reference to the wood duck and the mandarin duck. *Wilson Bull.*, 71: 46-53.

Dowling, J.E. and F.S. Werblin. 1969. Organization of the retina of the mudpuppy, *Necturus maculosus*, I. Synaptic structure. *J. Neurophysiol.*, 32: 315-338. (*146*)

Dwight, J., Jr. 1900. Sequence and plumage of moults of the passerine birds of New York. *Ann. N.Y. Acad. Sci.*, 13: 73-360. (*112*)

Dyke, J. 1971. Structure and colour production of the blue barbs of *Agapornis roseicollis* and *Cotinga maynana*. *Z. Zellforsch.*, 115: 17-29. (*101*)

Eaton, R.L. 1976. A possible case of mimicry in larger mammals. *Evolution*, 30: 853-856. (*177*)

Edmunds, M. 1974. *Defence in Animals: A Survey of Anti-predator Defences*. Harlow, Longman. 357 pp. (*189*)

Edmunds, M. 1976. The defensive behaviour of Ghanaian praying mantids with a discussion of territoriality. *Zool. J. Linn. Soc.*, 58: 1-37. (*189*)

Eibl-Eibesfeldt, I. 1970. *Ethology: The Biology of Behavior*. Holt, Rinehart and Winston, N.Y. 503 pp. (*286*)

Eisner, T., R.E. Silberglied, D. Aneshansley, J.E. Carrel and H.C. Howland. 1969. Ultraviolet video-viewing: the television camera as an insect eye. *Science*, 146: 1172-1174. (*170, 282*)

Emory, G.R. 1976. Aspects of attention, orientation and status hierarchy in mandrils (*Mandrillus sphinx*) and gelada baboons (*Theropithecus galada*). *Behaviour*, 59: 70-89. (*289*)

Ficken, R.W. and L.B. Wilmot. 1968. Do facial eye-stripes function in avian vision? *Amer. Midl. Nat., 79:* 522-523. (*114*)

Ficken, R.W., P.E. Matthiae and R. Horwich. 1971. Eye marks in vertebrates: aids to vision. *Science, 173:* 936-939. (*114*)

Fingerman, M. 1963. *The Control of Chromatophores.* MacMillan, N.Y. 184 pp. (*115*)

Firth, C.B. 1970. Sympatry of *Amblyornis subalaris* and *A. macgregoriae* in New Guinea. *Emu, 70:* 196-197. (*201*)

Ford, E.B. 1964. *Ecological Genetics.* John Wiley & Sons, N.Y. (*170*)

Fox, H.M. and G. Vevers. 1960. *The Nature of Animal Colours.* Sidgwick and Jackson, London. 246 pp. (*96, 115*)

Fox, D.L. 1962. Metabolic fractionation, storage and display of carotenoid pigments by flamingoes. *Comp. Biochem. Physiol., 6:* 1-40. (*112*)

Fox, D.L. 1953. *Animal Biochromes and Structural Colours.* Cambridge Univ. Press. 379 pp. (*95, 96, 115*)

Franq, E.N. 1969. Behavioral aspects of feigned death in the opossum *Didelphia marsupialis. Amer. Midl. Nat., 81:* 556-568. (*180*)

Friedmann, H. 1960. The parasitic weaverbirds. *Bull. U.S. Nat. Mus., 223:* 1-196. (*292*)

Frisch, K. von. 1955. *The Dancing Bees.* Harcourt, Brace, N.Y. (*31*)

Fuortes, M.G.F. (ed). 1972. *Handbook of Sensory Physiology, Vol. VII/2. Physiology of Photoreceptor Organs.* Springer-Verlag, N.Y. 765 pp. (*154*)

Gilliard, E.T. 1969. *Birds of Paradise and Bower Birds.* Natural History Press, Garden City, N.Y. (*205*)

Golani, I. 1976. Homeostatic motor processes in mammalian interactions --a choreography of display. *In:* P.P.G. Bateson and P.H. Klopfer (eds), *Perspectives in Ethology,* vol. 2, Plenum Press, N.Y. and London, pp. 69-134. (*264*)

Goldsmith, T.H. 1961. The color vision of insects. *In:* W.D. McElroy and B. Glass (eds), *Light and Life,* Johns Hopkins Press, Baltimore, pp. 771-794. (*135, 136-137*).

Goodwin, T.W. 1954. *Carotenoids.* Chem. Publ. Co., N.Y. (*115*)

Gould, J.L. 1975. Honey bee recruitment: the dance-language controversy. *Science, 189:* 685-691. (*32*)

Grand, Y. Le. 1957. *Light, Colour and Vision.* Chapman & Hall, London. (*155*)

Gregory, R.L. 1972. Cognitive contours. *Nature, 238:* 51-52. (*145*)

Grubb, R.C., Jr. 1971. Bald eagles stealing fish from common mergansers. *Auk, 88:* 928-929. (*158*)

Guthrie, R.D. and R.G. Petocz. 1970. Weapon automimicry among mammals. *Amer. Nat., 104:* 585-588. (*308*)

Haartman, L. von. 1957. Adaptation in hole-nesting birds. *Evolution, 11:* 339-347. (*252, 291*)

Hailman, J.P. 1958. Notes on pre-copulatory display in the starling. *Wilson Bull., 70:* 199-201. (*290*)

Hailman, J.P. 1959a. Unusual "bunching" behavior of starlings. *Condor, 61:* 369. (*297*)

Hailman, J.P. 1959b. Why is the male wood duck strikingly colorful? *Amer. Nat., 93:* 383-384. (*312*)

Hailman, J.P. 1963. Why is the Galapagos lava gull the color of lava? *Condor, 65:* 528. (*158*)

Hailman, J.P. 1965. Cliff-nesting adaptations of the Galapagos swallow-tailed gull. *Wilson Bull., 77:* 346-362. *(15)*

Hailman, J.P. 1967a. The ontogeny of an instinct. *Behav. Suppl., 15:* 1-196. *(12, 194, 206, 263, 312)*

Hailman, J.P. 1967b. Spectral discrimination: an important correction. *J. Opt. Soc. Amer., 57:* 281-282. *(133)*

Hailman, J.P. 1970. Comments on the coding of releasing stimuli. *In:* L.R. Aronson, E. Tobach, D.S. Lehrman and J.S. Rosenblatt (eds), *Development and Evolution of Behavior,* W.H. Freeman, San Francisco, pp. 138-157. *(145)*

Hailman, J.P. 1971. The role of stimulus-orientation in eliciting the begging response from newly-hatched chicks of the laughing gull *(Larus atricilla). Anim. Behav., 19:* 328-335. *(199, 312)*

Hailman, J.P. 1975. The scientific method: *modus operandi* or supreme court? *Amer. Biol. Teacher, 37:* 309-310. *(9)*

Hailman, J.P. 1976a. Uses of the comparative study of behavior. Chapter 2 *in:* R.B. Masterton, W. Hodos and H. Jerison (eds), *Evolution, Brain, and Behavior: Persistent Problems.* Erlbaum, Hillsdale, N.J. pp. 181-189. *(11, 12, 292)*

Hailman, J.P. 1976b. L'ontogenese du comportement. *La Recherche, 7:* 520-528. *(12)*

Hailman, J.P. 1976c. Oildroplets in the eyes of adult anuran amphibians: a comparative survey. *J. Morphol., 148:* 453-468. *(113)*

Hailman, J.P. 1977a. Communication by reflected light. Chapter 9 *in:* T.A. Sebeok (ed), *How Animals Communicate,* Indiana Univ. Press, Bloomington, *in press. (6, 12, 13, 27, 30, 33, 38, 48, 91, 102, 103, 104, 105, 106, 107, 109, 110, 115, 149, 193, 195, 199, 200, 252, 253, 256, 257, 260, 264, 268, 274, 275, 276, 277, 280, 286, 297, 304, 308, 311-312)*

Hailman, J.P. 1977b. Bee dancing and evolutionary epistemology. *Amer. Nat., in press. (12, 108)*

Hailman, J.P. and J.J.I. Dzelzkalns. 1974. Mallard tail-wagging: punctuation for animal communication? *Amer. Nat., 108:* 236-238. *(267)*

Hailman, J.P. and R.G. Jaeger, 1976. A model of phototaxis and its evaluation with anuran amphibians. *Behaviour, 56:* 215-249. *(122, 123)*

Hailman, J.P. and B.D. Sustare. 1973. What a stuffed toy tells a stuffed shirt. *BioScience, 23:* 644-651. *(37)*

Haldane, J.B.S. and H. Spurway. 1954. A statistical analysis of communication in *"Apis mellifera"* and a comparison with communication in other animals. *Insectes Sociaux, 1:* 247-283. *(31-32)*

Hall, K.R.L. and I. DeVore. 1965. Baboon social behavior. *In:* I. DeVore (ed), *Primate Behavior: Field Studies of Monkeys and Apes,* Holt, Rinehart and Winston, N.Y., pp. 53-110. *(251)*

Hamilton, T.H. 1961. On the functions and causes of sexual dimorphism in breeding plumage characters of North American species of warblers and orioles. *Amer. Nat., 95:* 121-123. *(280, 289)*

Hamilton, T.H. and R.H. Barth, Jr. 1962. The biological significance of seasonal change in male plumage appearance in some New World migratory bird species. *Amer. Nat., 94:* 129-144. *(281)*

Hamilton, W.J. III. 1965. Sun-oriented display of the Anna's hummingbird. *Wilson Bull., 77:* 38-43. *(217, 241)*

Hamilton, W.J., III. 1973. *Life's Color Code.* McGraw-Hill, N.Y. 238 pp. *(110, 111)*

Hamilton, W.J., III and R.M. Peterman. 1971. Countershading in the colourful reef fish *Chaetodon lunula:* concealment, communication or both? *Anim. Behav., 19:* 357-364. *(196-197)*

Harrison, C.J.O. 1965. Allopreening as agonistic behaviour. *Behaviour*, *24:* 161-209. (*294*)

Hartshorne, C. and P. Weiss (eds). 1931-1935. *The Collected Papers of Charles Sanders Peirce*, Vols. 1-6; [Burks, A.W. (ed), 1958, Vols. 7-8]; Harvard Univ. Press, Cambridge, Mass. (*53*)

Harvey, E.N. 1960. Bioluminescence. *In:* M. Florkin and H.S. Mason (eds), *Comparative Biochemistry*, Vol. II, pp. 545-591, Academic Press, N.Y. (*115*)

Harvey, E.N. 1952. *Bioluminescence.* Academic Press, N.Y. 649 pp. (*115*)

Hastings, J.W. 1971. Light to hide by: ventral luminescence to camouflage the silhouette. *Science*, *173:* 1015-1017. (*163*)

Heaton, M.B. and M.S. Harth. 1974. Non-visual light responsiveness in the pigeon: developmental and comparative considerations. *J. Exp. Zool.*, *188:* 251-264. (*120*)

Hinde, R.A. 1970. *Animal Behaviour: A Synthesis of Ethology and Comparative Psychology*, 2nd ed. McGraw-Hill, N.Y. and elsewhere. 876 pp. (*53*)

Hingston, R.W.G. 1933. *The Meaning of Animal Colour and Adornment.* Edward Arnold, London. (*292*)

Hoar, W.S. 1975. *General and Comparative Physiology*, 2nd ed. Prentice-Hall, Englewood Cliffs, N.J. 848 pp. (*115*)

Hogan-Warburg, A.J. 1966. Social behaviour of the ruff, *Philomachus pugnax. Ardea*, *54:* 109-229. (*277*)

Hubel, D.H. and T.N. Wiesel. 1959. Receptive fields of single neurons in the cat's striate cortex. *J. Physiol.*, *148:* 574-591. (*147*)

Hubel, D.H. and T.N. Wiesel. 1960. Receptive fields of optic nerve fibers in the spider monkey. *J. physiol.*, *154:* 572-580. (*148*)

Hubel, D.H. and T.N. Wiesel. 1961. Integrative action in the cat's lateral geniculate body. *J. Physiol.*, *155:* 385-398. (*147*)

Hubel, D.H. and T.N. Wiesel. 1962. Receptive fields, binocular interaction and functional architecture in the cat's visual cortex. *J. Physiol.*, *160:* 106-154. (*147*)

Hubel, D.H. and T.N. Wiesel. 1963. Shape and arrangement of columns in the cat's striate cortex. *J. Physiol.*, *165:* 559-568. (*147*)

Hubel, D.H. and T.N. Wiesel. 1965. Receptive fields and functional architecture in two nonstriate visual areas (18 and 19) of the cat. *J. Neurophysiol.*, *28:* 229-289. (*147*)

Huey, R.B. and E.R. Pianka. 1977. Natural selection for juvenile lizards mimicking noxious beetles. *Science*, *195:* 201-203. (*177*)

Hulst, H.C. van de. 1957. *Light Scattering by Small Particles.* Wiley & Sons, N.Y. 470 pp. (*88*)

Hunsaker, D. 1962. Ethological isolating mechanisms in the *Sceloporus torquatus* group of lizards. *Evolution*, *16:* 62-74. (*280*)

Hurvich, L.M. and D. Jameson. 1966. *The Perception of Brightness and Darkness.* Allyn & Bacon, Boston. 141 pp. (*123*)

Hutchinson, G.E. 1957. *A Treatise on Limnology*, Vol. I. Wiley, N.Y. (*84*)

Hutchinson, G.E. and R.H. MacArthur. 1959. Appendix: on the theoretical significance of aggressive neglect in interspecific competition. *Amer. Nat.*, *93:* 133-134. (*284*)

Huxley, J.S. 1914. The courtship habits of the great crested grebe (*Podiceps cristatus*): with an addition to the theory of sexual selection. *Proc. Zool. Soc. London*, *35:* 491-562. (*285, 286*)

Huxley, J.S. 1923. Courtship activities in the red-throated diver

(*Colymbus stellatus* Pontopp.); together with a discussion of the evolution of courtship in birds. *J. Linn. Soc. London, Zool., 35:* 253-292. *(24)*

Huxley, J.S. 1958. Why two breast-bands on the killdeer? *Auk, 75:* 98-99. *(166)*

Jackson, J.F., W. Ingram, III and H.W. Campbell. 1976. The dorsal pigmentation pattern of snakes as an antipredator strategy: a multivariate approach. *Amer. Nat., 110:* 1029-1053. *(185)*

Jacobson, M. and R.M. Gaze. 1964. Types of visual response from single units in the optic tectum and optic nerve of the goldfish. *Quart. J. Exp. Physiol., 49:* 199-209. *(149)*

Jenssen, T.A. and E.L. Hover. 1976. Display analysis of the signature display of *Anolis limifrons* (Sauria: Iguanidae). *Behaviour, 57:* 227-240. *(263, 267)*

Jerlov, N.G. 1968. *Optical Oceanography.* Elsevier, Amsterdam. 194 pp. *(84)*

Johnson, F.H. and Y. Haneda (eds). 1966. *Bioluminescence in Progress.* Princeton Univ. Press, Princeton, N.J. *(115)*

Johnson, N.K. and A.H. Brush. 1972. Analysis of polymorphism in the sooty-capped bush tanager. *System. Zool., 21:* 245-262. *(170)*

Johnston, T.D. 1976. Theoretical considerations in the adaptation of animal communication systems. *J. Theor. Biol., 57:* 43-72. *(26, 27, 53, 233, 262, 307)*

Jouventin, P. 1975. Les roles des colorations du mandrill (*Mandrillus sphinx*). *Z. Tierpsychol., 39:* 455-462. *(258)*

Karrer, P. and E. Jucker. 1950. *Carotenoids.* Elsevier, N.Y. *(115)*

Keenleyside, M.H.A. 1955. Some aspects of the schooling behaviour of fish. *Behaviour, 8:* 183-248. *(295, 296)*

Keenleyside, M.H.A. 1972. The behaviour of *Abudefduf zonatus* (Pisces, Pomacentridae). *Anim. Behav., 20:* 763-774. *(196, 282)*

Kessel, E.L. 1955. Mating activities of balloon flies. *System. Zool., 4:* 97-104. *(258)*

Kettlewell, B. 1973. *The Evolution of Melanism.* Oxford Univ. Press. *(170)*

Kleiman, D.G. and J.F. Eisenberg. 1973. Comparison of canid and felid social systems from an evolutionary perspective. *Anim. Behav., 21:* 637-659. *(278)*

Klopfer, P.H. 1974. *An Introduction to Animal Behavior,* 2nd ed. Prentice-Hall, Englewood Cliffs, N.J. 332 pp. *(53)*

Klopfer, P.H. and J.P. Hailman. 1967. *An Introduction to Animal Behavior.* Prentice-Hall, Englewood Cliffs, N.J. 297 pp. *(317)*

Klopfer, P.H. and J.J. Hatch. 1968. Experimental considerations. *In:* T.A. Sebeok (ed), *Animal Communication,* pp. 31-43, Indiana Univ. Press, Bloomington. *(24, 47)*

Klots, A.B. 1951. *A Field Guide to the Butterflies of North America, East of the Great Plains.* Houghton Mifflin, Boston. 349 pp. *(175, 182)*

Konishi, M. 1970. Evolution of design features in the coding of species-specificity. *Amer. Zool., 10:* 67-72. *(1)*

Kreithen, M.L. and W.T. Keeton. 1974. Detection of polarized light by the homing pigeon, *Columba livia. J. Comp. Physiol., 89:* 83-92. *(120)*

Kuffler, S.W. 1953. Discharge patterns and functional organization of mammalian retina. *J. Neurophysiol., 16:* 37-68. *(146)*

Kühme, W. 1963. Ergänzende Beobachtungen an afrikanischen Elefanten (*Loxodonta africana* Blumenbach 1797) in Freigehege. *Z. Tierpsychol.*, *20:* 66-79. (*277*)

Kummer, H. 1968. Social organization of Hamadryas baboons. *Biblio. Primat.*, *6:* 1-189. (*295*)

Lack, D. 1940. Courtship feeding in birds. *Auk*, *57:* 169-178. (*287*, *289*)

Lawick-Goodall, J. van. 1968. The behaviour of free-living chimpanzees in the Gombe Stream Reserve. *Anim. Behav. Monogr.*, *1:* 161-301. (*290*)

Lawson, R.B. and W.L. Gulick. 1967. Stereopsis and anomalous contour. *Vision Res.*, *7:* 271-297. (*145*)

Lettvin, J.Y., H.R. Maturana, W.H. Pitts and W.S. McCulloch. 1961. Two remarks on the visual system of the frog. *In:* W.A. Rosenblith (ed), *Sensory Communication,* pp. 757-776, M.I.T. Press, Cambridge, Mass. (*148*)

Leyhausen, P. 1956. Verhaltensstudien bei Katzen. *Z. Tierpsychol. Beiheft*, *2:* 1-100. (*269*)

Lieb, I.C. (ed). 1953. *Charles S. Peirce's Letters to Lady Welby.* Whitlock, New Haven, Conn. (*53*)

Liebman, P.A. and G. Entine. 1968. Visual pigments of frog and tadpole *(Rana pipiens). Vision Res.*, *8:* 761-775. (*118*)

Linsemair, K.E. 1967. Konstruktion und Signalfunktion der Sandpyramide der Reiterkrabbe *Ocypode saratan* Forsk. (Decapoda, Brachyura, Ocypodidae). *Z. Tierpsychol.*, *24:* 403-456. (*283*)

Lloyd, J.E. 1977. Bioluminescence and communication. *In:* T.A. Sebeok (ed), *How Animals Communicate,* Indiana Univ. Press, Bloomington, *in press.* (*93, 106, 115, 163, 178*)

Loomis, W.F. 1967. Skin-pigmentation regulation of vitamin D biosynthesis in man. *Science*, *157:* 501-506. (*111*)

Lorenz, K.Z. 1941. Vergleichende Bewegungsstudien an Anatinen. *J. Ornithol.*, *89:* 194-293. (*240, 253, 266*)

Lorenz, K.Z. 1950. The comparative method in studying innate behaviour patterns. *In:* S.E.B. Symposium, No. 4: *Physiological Mechanisms in Animal Behaviour,* pp. 221-268, Academic Press, N.Y. (*34*)

Lorenz, K.Z. 1952. *King Solomon's Ring: New Light on Animal Ways.* Crowell, N.Y. 202 pp. (*278*)

Luria, S.M. and J.A. Kinney. 1970. Underwater vision. *Science*, *167:* 1454-1461. (*84*)

MacArthur, R.H. 1958. Population ecology of some warblers of northeastern coniferous forests. *Ecology*, *39:* 599-619. (*232*)

Magnuson, J.J. and R.M. Gooding. 1971. Color patterns of pilotfish *(Naucrates ductor)* and their possible significance. *Copeia*, *1971:* 314-316.

Makkink, G.F. 1936. An attempt at an ethogram of the European avocet with ethological and psychological remarks. *Ardea*, *25:* 1-60. (*277*, *278*)

Mandelbrot, B. 1953. Contribution a la theorie mathematique des jeux de communication. *Publ. l'Inst. Stat. l'Univ. Paris*, *2:* 5-50. (*262*)

Marks, H.L., P.B. Siegel and C.Y. Kramer. 1960. Effect of comb and wattle removal on the social organization of mixed flocks of chickens. *Anim. Behav.*, *8:* 192-196. (*279*)

Marks, W.B. 1965. Visual pigments of single goldfish cones. *J. Phys-*

iol., 178: 14-32. (*131*)

Marler, P. 1961. The logical analysis of animal communication. *J. Theor. Biol., 1:* 295-317. (*27, 49, 249*)

Marler, P. 1968. Visual system. *In:* T.A. Sebeok (ed), *Animal Communication,* pp. 103-126, Indiana Univ. Press, Bloomington. (*23, 24, 26*)

Marler, P. 1974. Animal communication. Chapter 2 *in:* L. Krames, P. Pliner and T. Alloway (eds), *NonVerbal Communication,* pp. 25-50, Plenum, N.Y. (*195*)

Marler, P. 1977. The evolution of communication. Chapter 5 *in:* T.A. Sebeok (ed), *How Animals Communicate,* Indiana Univ. Press, Bloomington, *in press.* (*256*)

Marler, P. and W.J. Hamilton, III. 1966. *Mechanisms of Animal Behavior.* Wiley, N.Y. (*300*)

Marshall, A.J. 1953. *Bower-birds: Their Displays and Breeding Cycles.* Clarendon Press, Oxford. (*205*)

Maturana, H.R. and S. Frenk. 1963. Directional movement and horizontal edge detectors in the pigeon retina. *Science, 142:* 977-978. (*147*)

Maturana, H.R., J.Y. Lettvin, W.S. McCulloch and W.H. Pitts. 1960. Anatomy and physiology of vision in the frog. *J. Gen. Physiol., 43:* 127-177. (*147*)

Maynard Smith, J. 1974. The theory of games and the evolution of animal conflicts. *J. Theor. Biol., 47:* 209-221. (*274*)

Mayr, E. 1963. *Animal Species and Evolution.* Harvard Univ. Press, Cambridge, Mass. (*280, 307*)

McCullough, E.C. and W.P. Porter. 1971. Computing clear day solar radiation spectra for the terrestrial ecological environment. *Ecology, 52:* 1008-1015. (*75, 76*)

McFarland, W.N. and F.W. Munz. 1975a. Part II. The photic environment of clear tropical seas during the day. *Vision Res., 15:* 1063-1070. (*81, 82, 83*)

McFarland, W.N. and F.W. Munz. 1975b. Part III. The evolution of photopic visual pigments in fishes. *Vision Res., 15:* 1071-1080. (*118*)

McKinney, F. 1975. The evolution of duck displays. Chapter 16 *in:* G. Baerends, C. Beer and A. Manning (eds), *Function and Evolution in Behaviour,* pp. 331-357, Oxford Univ. Press. (*266, 287, 291, 298*)

Menaker, M. 1968. Extraretinal light perception in the sparrow. I. Entrainment of the biological clock. *Proc. Natl. Acad. Sci., 59:* 414-421. (*111*)

Menaker, M., R. Roberts, J. Elliott and H. Underwood. 1970. Extraretinal light perception in the sparrow. III. The eyes do not participate in photoperiod photoreception. *Proc. Natl. Acad. Sci., 67:* 320-325. (*120*)

Meyerriecks, A.J. 1960. Comparative breeding behavior of four species of North American herons. *Publ. Nuttall Ornithol. Club, 2:* 1-158. (*227, 288*)

Michael, C.R. 1968. Receptive fields of single optic nerve fibers in a mammal with an all-cone retina. I. Contrast-sensitive units. *J. Neurophysiol., 31:* 249-256. (*149*)

Miller, D.J., D. Gotshall and R. Nitsos. 1965. *A Field Guide to some Common Ocean Sport Fishes of California.* Calif. Dept. Fish Game. 87 pp. (*171*)

Miller, E.H. 1975. Walrus ethology. I. The social role of tusks and applications of multidimensional scaling. *Canad. J. Zool., 53:* 590-613. (*275*)

Minnaert, M. 1954. *The Nature of Light and Color in the Open Air*. Dover Publ., N.Y. 362 pp. (*88, 123, 149, 231, 276*)

Moles, A.A. 1968. Perspectives for communication theory. Chapter 23 *in:* T.A. Sebeok (ed), *Animal Communication*, pp. 627-642, Indiana Univ. Press, Bloomington. (*262*)

Monk, G.S. 1963. *Light: Principles and Experiments*, 2nd ed. Dover, N.Y. 489 pp. (*88*)

Moody, M.I. 1975. Perception of total reflection by *Barbus*. *Behav. Biol., 15:* 239-243. (*66*)

Morin, J.G., H. Harrington, K. Nealson, N. Krieger, T.O. Baldwin and J.W. Hastings. 1975. Light for all reasons: versatility in the behavioral repertoire of the flashlight fish. *Science, 190:* 74-76. (*183*)

Morris, C.W. 1946. *Signs, Language and Behavior*. Braziller, N.Y. (*27, 48-49, 53, 249*)

Morris, D. 1954. The reproductive behaviour of the zebra finch *(Poephila guttata)* with special reference to pseudofemale behaviour and displacement activities. *Behaviour, 6:* 271-322. (*283*)

Morris, D. 1957. "Typical intensity" and its relation to the problem of ritualization. *Behaviour, 11:* 1-12. (*268*)

Morris, D. 1958. The reproductive behaviour of the ten-spined stickleback (*Pygosteus pungitius* L.). *Behav. Suppl., 6:* 1-154. (*196*)

Morse, D.H. 1970. Ecological aspects of some mixed species foraging flocks of birds. *Ecol. Monogr., 40:* 119-168. (*297*)

Moynihan, M. 1960. Some adaptations which help to promote gregariousness. *Proc. XII Internat. Ornithol. Congr.*, Helskinki, pp. 523-541. (*281*)

Moynihan, M. 1962. The organization and probable evolution of some mixed-species flocks of neotropical birds. *Smithsonian Misc. Collect., 143:* 1-140. (*297*)

Moynihan, M. 1968. Social mimicry: character convergence versus character displacement. *Evolution, 22:* 315-331. (*295, 305*)

Moynihan, M. 1970. Control, suppression, decay, disappearance and replacement of displays. *J. Theor. Biol., 29:* 85-112. (*34, 207-208, 262, 263*)

Moynihan, M. 1975. Conservatism of displays and comparable stereotyped patterns among cephalopods. Chapter 13 *in:* G. Baerends, C. Beer and A. Manning (eds), *Function and Evolution in Behaviour*, pp. 276-291, Oxford Univ. Press. (*159, 175, 176, 186, 238*)

Munz, F.W. and S.A. Schwanzara. 1967. A nomogram for retinene$_2$-based visual pigments. *Vision Res., 7:* 111-120. (*118*)

Needham, A.E. 1974. *The Significance of Zoochromes*. Springer-Verlag, Berlin. 429 pp. (*95, 96, 112, 113-114, 115, 120*)

Nelson, K. 1964. The temporal patterning of courtship behavior in the glandulocaudine fishes (Ostariophysi, Characidae). *Behaviour, 24:* 90-146. (*254*)

Nero, R.W. 1956. A behavior study of the red-winged blackbird. *Wilson Bull., 68:* 5-37, 129-150. (*4*)

Nisbet, I.C.T. 1973. Courtship-feeding, egg size and breeding success in common terns. *Nature, 241:* 141-142. (*287*)

Noble, G.K. 1936. Courtship and sexual selection of the flicker *(Colaptes auratus leuteus)*. *Auk, 53:* 269-282. (*27*)

Noble, G.K. and B. Curtis. 1939. The social behavior of the jewel fish,

Hemichromis bimaculatus Gill. *Bull. Amer. Mus. Nat. Hist., 76:* 1-46. (*195*)

Orians, G.H. 1969. On the evolution of mating systems in birds and mammals. *Amer. Nat., 103:* 589-603. (*285*)

Otte, D. 1975. On the role of intraspecific deception. *Amer. Nat., 109:* 239-242. (*270*)

Owen, D.F. 1961. Industrial melanism in North American moths. *Amer. Nat., 95:* 227-233. (*170*)

Parker, G.H. 1948. *Animal Colour Changes and their Neurohumours.* Cambridge Univ. Press, London & N.Y. 377 pp. (*115*)

Parker, G.A. 1974. Assessment strategy and the evolution of fighting behaviour. *J. Theoret. Biol., 47:* 223-243. (*274*)

Peterson, R.T. and E.L. Chalif. 1973. *A Field Guide to Mexican Birds.* Houghton Mifflin, Boston. 298 pp. (*281*)

Pierce, J.R. 1961. *Symbols, Signals and Noise.* Harper & Bros. (*53*)

Pietrewicz, A.T. and A.C. Kamil. 1977. Visual detection of cryptic prey by blue jays *(Cyanocitta cristata). Science, 195:* 580-582. (*159*)

Pijl, L. van der and C.H. Dodson. 1966. *Orchid Flowers: Their Pollination and Evolution.* Univ. Miami Press, Coral Gables, Fla. 214 pp. (*202*)

Pirenne, M.H. 1967. *Vision and the Eye,* 2nd ed. Chapman & Hall, London. 224 pp. (*142, 155*)

Pitcher, T.J., B.L. Partridge and C.S. Wardle. 1976. A blind fish can school. *Science, 194:* 963-965. (*294*)

Popper, K.R. 1959. *The Logic of Scientific Discovery.* Harper & Row, N.Y. (*18*)

Porter, W.P. 1967. Solar radiation through the living body walls of vertebrates with emphasis on desert reptiles. *Ecol. Monogr., 37:* 273-296. (*110*)

Porter, W.P., J.W. Mitchell, W.A. Beckman and C.B. DeWitt. 1973. Behavioral implications of mechanistic ecology: thermal and behavioral modeling of desert ectotherms and their microenvironment. *Oecologia, 13:* 1-54. (*110-111*)

Prosser, C.L. and F.A. Brown, Jr. 1961. *Comparative Animal Physiology,* 2nd ed. Saunders, Philadelphia. 688 pp. (*115*)

Rand, A.S. and E.E. Williams. 1970. An estimation of redundancy and information content of anole dewlaps. *Amer. Nat., 104:* 99-103. (*52*)

Raven, P.P. 1972. Why are bird-visited flowers predominantly red? *Evolution, 26:* 674. (*202*)

Rhijn, J.G. van. 1974. Behavioural dimorphism in male ruffs. *Behaviour, 97:* 153-229. (*277*)

Rijke, A.M. 1968. The water repellency and feather structure of cormorants, Phalacrocoracidae. *J. Exp. Biol., 48:* 185-189. (*109*)

Robbins, C.S., B. Bruun and H.S. Zim. 1966. *Birds of North America.* Golden Press, N.Y. 340 pp. (*243*)

Robinson, N. 1966. *Solar Radiation.* Elsevier Press, N.Y. (*88*)

Rodieck, R.W. 1973. *The Vertebrate Retina: Principles of Structure and Function.* W.H. Freeman, San Francisco. 1044 pp. (*154*)

Ruechardt, E. 1958. *Light Visible and Invisible.* Univ. Mich Press, Ann Arbor. 201 pp. (*76, 167*)

Ruiter, L. de. 1958. Some remarks on problems of the ecology and evolution of mimicry. *Arch. Neerl. Zool.*, *13*, Suppl. 1: 351-368. *(161)*

Russell, B. 1945. *A History of Western Philosophy*. Simon & Schuster, N.Y. *(18)*

Sackett, G.P. 1966. Monkeys reared in isolation with pictures as visual input: evidence for an innate releasing mechanism. *Science, 154:* 1468, 1471-1473. *(311, 312)*

Schauensee, R.M. de. 1970. *A Guide to the Birds of South America*. Livingston, Wynnewood, Pa. 470 pp. *(281)*

Schenkel, R. 1948. Ausdrucksstudien an Wolfen. *Behaviour, 1:* 81-130. *(278)*

Schleidt, W.M. 1961. Reaktionen von Truthühern auf fliegende Raubvögel und Versuche zur Analyse ihrer AAM's. *Z. Tierpsychol., 18:* 534-560. *(194)*

Schleidt, W.M. 1964. Ueber die Spontaneität von Erbkoordinationen. *Z. Tierpsychol., 21:* 235-256. *(194)*

Schleidt, W.M. 1973. Tonic communication: continual effects of discrete signs in animal communication systems. *J. Theor. Biol., 42:* 359-386. *(45-46, 47, 191, 262, 263, 267, 268)*

Schoener, T.W. and A. Schoener. 1976. The ecological context of female pattern polymorphism in the lizard *Anolis sagrei*. *Evolution, 30:* 650-658. *(171)*

Schrier, A.M. and D.S. Blough. 1966. Photopic spectral sensitivity of macaque monkeys. *J. Comp. Physiol. Psychol., 62:* 457-458. *(129)*

Schusterman, R.J. and B. Burrett. 1973. Amphibious nature of visual acuity in the Asian "clawless" otter. *Nature, 244:* 518-519. *(143)*

Sebeok, T.A. 1965. Animal communication. *Science, 147:* 1006-1014. *(249)*

Sebeok, T.A. (ed). 1968. *Animal Communication*. Indiana Univ. Press, Bloomington. 686 pp. *(26, 53, 249)*

Sebeok, T.A. 1975. Notes on lying and prevarication. *Rev. Roumaine Linguist., 20:* 571-574. *(157)*

Sebeok, T.A. (ed). 1977. *How Animals Communicate*. Indiana Univ. Press, Bloomington. *In press*. *(53, 300)*

Selander, R.K. and C.J. La Rue, Jr. 1961. Interspecific preening invitation display of parasitic cowbirds. *Auk, 78:* 473-504. *(294)*

Shannon, C.E. and W. Weaver. 1949. *The Mathematical Theory of Communication*. Univ. Illinois Press, Urbana. 117 pp. *(29, 31, 49-50, 51, 53, 55, 191, 249)*

Shaw, E. 1970. Schooling in fishes: critique and review. *In:* L.R. Aronson, E. Tobach, D.S. Lehrman and J.S. Rosenblatt (eds), *Development and Evolution of Behavior*, pp. 452-480, Freeman, San Francisco. *(295)*

Sibley, C. 1957. The evolutionary and taxonomic significance of sexual dimorphism and hybridization in birds. *Condor, 59:* 166-191. *(280)*

Simmons, K.E.L. and U. Weidmann. 1973. Directional bias as a component of social behaviour with special reference to the mallard, *Anas platyrhynchos*. *J. Zool, Lond., 170:* 49-62. *(266)*

Simpson, M.J.A. 1973. Social displays and the recognition of individuals. Chapter 7 *in:* P.P.G. Bateson and P.H. Klopfer (eds), *Perspectives in Ethology*, pp. 225-279, Plenum, N.Y. *(195, 207, 298)*

Singh, J. 1966. *Great Ideas in Information Theory, Language and Cybernetics*. Dover, N.Y. 338 pp. *(53)*

Smith, W.J. 1968. Message-meaning analyses. *In:* T.A. Sebeok (ed), *Animal Communication*, pp. 44-60, Indiana Univ. Press, Bloomington.

(23, 24, 26, 27, 44, 49, 249)

Smith, W.J. 1969. Messages of vertebrate communication. *Science, 165:* 145-150. (262)

Smith, W.J. 1977. *The Behavior of Communicating: An Ethological Approach.* Harvard Univ. Press, Cambridge, Mass. *In press.* (300)

Sparks, J.H. 1965. On the role of allopreening invitation behaviour in reducing aggression among red avadavants, with comments on its evolution in the Spermestidae. *Proc. Zool. Soc. London, 145:* 387-403. (294)

Stephenson, G.R. 1973. Testing for group specific communication patterns in Japanese macaques. *In:* W. Montagna and E.W. Menzel, Jr. (eds), *Symposia of the IVth Inter. Congr. Primatol., vol. I: Precultural Primate Behavior,* pp. 51-75, Karger, Basel. (27, 311)

Sterba, G. 1962. *Freshwater Fishes of the World.* Pet Library, N.Y. 877 pp. (162)

Steven, D.M. 1963. The dermal light sense. *Biol. Rev., 38:* 204-240. (120)

Struhsaker, T.T. 1967. Auditory communication among vervet monkeys *(Cercopithecus aethiops).* *In:* S.A. Altmann (ed), *Social Communication Among Primates,* pp. 281-324. (251)

Sustare, B.D. 1976. Comparative electroretinography of anuran amphibians. Ph.D. Thesis, Univ. Wisconsin-Madison. 99 pp. (121, 129)

Tavolga, W.N. 1968. Fishes. *In:* T.A. Sebeok (ed), *Animal Communication,* pp. 271-288, Indiana Univ. Press, Bloomington. (24)

Tembrock, G. 1968. Land mammals. Chapter 16 *in:* T.A. Sebeok (ed), *Animal Communication,* pp. 338-404, Indiana Univ. Press, Bloomington. (265, 269)

Test, F.H. 1942. The nature of the red, yellow, and orange pigments in woodpeckers of the genus *Colaptes.* *Univ. Calif. Publ. Zool., 46:* 371-390. (100)

Thayer, G.H. 1918. *Concealing-coloration in the Animal Kingdom.* Mac-Millan, N.Y. (161)

Thekaekara, M.P. 1972. Evaluating the light from the sun. *Optical Spectra,* March 1972: 32-35. (75)

Tinbergen, L. 1946. Der Sperwer als Roofvijand van Zangvogels. *Ardea, 34:* 1-213. (184)

Tinbergen, N. 1951. *The Study of Instinct.* Oxford Univ. Press, London. 228 pp. (34, 53, 106, 110, 237, 241, 275, 283, 300)

Tinbergen, N. 1959. Comparative studies of the behaviour of gulls (Laridae): a progress report. *Behaviour, 15:* 1-70. (15, 265, 275, 289, 290)

Tinbergen, N. 1963. On aims and methods of ethology. *Z. Tierpsychol., 20:* 410-429. (11)

Tyler, J.E. 1960. Radiance distribution as a function of depth in an underwater environment. *Bull. Scripps Inst. Oceanogr., 7:* 363-412. (83)

Verner, J. 1964. Evolution of polygyny in the long-billed marsh wren. *Evolution, 18:* 252-261. (285)

Walker, M. 1963. *The Nature of Scientific Thought.* Prentice-Hall, Englewood Cliffs, N.J. (6)

Wallace, B. 1973. Misinformation, fitness and selection. *Amer. Nat., 107:* 1-7. (270)

Walsh, J.W.T. 1958. *Photometry.* Dover, N.Y. 544 pp. (155)

Walther, F. 1964. Einige Verhaltensbeobachtungen an Thomsongazellen (*Gazella thomsoni* Gunther, 1884) im Ngorongoro-Krater. *Z. Tier-psychol., 21:* 871-890. (296)

Waterman, T.H. and K.W. Horch. 1966. Mechanism of polarized light perception. *Science, 154:* 467-475. (120)

Weidmann, U. 1956. Verhaltensstudien an der Stockente (*Anas platy-rhynchos* L.). I. Das Aktionssystem. *Z. Tierpsychol., 13:* 208-271. (291)

Werblin, F.S. 1971. Adaptation in a vertebrate retina: intracellular recording in *Necturus*. *J. Neurophysiol., 34:* 228-241. (146)

Werblin, F.S. 1972. Lateral interactions at inner plexiform layer of vertebrate retina: antagonistic responses to change. *Science, 175:* 1008-1010. (146)

Werblin, F.S. and J.E. Dowling. 1969. Organization of the retina of the mudpuppy, *Necturus maculosus*, II: intracellular recording. *J. Neurophysiol., 32:* 339-355. (146)

Wickler, W. 1968. *Mimicry in Plants and Animals*. McGraw-Hill, N.Y. 255 pp. (149, 157, 169, 171, 172, 174, 175, 176, 178, 181-182, 183, 184, 189, 255, 257, 278, 305, 308, 309)

Wiener, N. 1948. *Cybernetics*. Wiley & Sons, N.Y. (53)

Wiesel, T.N. and D.H. Hubel. 1966. Spatial and chromatic interactions in the lateral geniculate body of the rhesus monkey. *J. Neurophysiol., 29:* 1115-1156. (148)

Wiley, R.H. 1973. The strut display of male sage grouse: a "fixed" action pattern. *Behaviour, 47:* 129-152. (263)

Willson, M.F. and R.A. Neumann. 1972. Why are neotropical birds more colorful than North American birds? *Avicult. Mag., 78:* 141-147. (263)

Wilson, E.O. 1962. Chemical communication among workers of the fire ant *Solenopsis saevissima* (Fr. Smith). *Anim. Behav., 10:* 134-164. (32)

Wilson, E.O. 1968. Chemical systems. *In:* T.A. Sebeok (ed), *Animal Communication*, pp. 75-102, Indiana Univ. Press, Bloomington. (24)

Wilson, E.O. 1972. Animal communication. *Sci. Amer., 227*(3): 52-60. (263, 286)

Wilson, E.O. 1975. *Sociobiology*. Harvard Univ. Press, Cambridge, Mass. 697 pp. (12, 103, 271, 273, 300)

Wooton, R.J. 1971. Measures of the aggression of parental male three-spined sticklebacks. *Behaviour, 40:* 228-262. (231)

Zipf, G.K. 1935. *The Psycho-biology of Language*. Houghton Mifflin, Boston. 336 pp. (262)

INDEXES

Index to Persons

Authors of published works cited in the terminal bibliography are referenced here by the page numbers on which their works appear in the *LITERATURE CITED AND INDEXED;* in that section, all text-citations of each reference are listed. Personal communications, works in preparation and other citations of names appearing in the text, as well as authors of articles listed in the *APPENDIX*, are all referenced here according to the page on which the name appears.

Index to Animals

The index divides animals into the major taxonomic classes dealt with in the text: mammals, birds, reptiles and amphibians (combined), fishes (Osteichthyes and Chondrichthyes combined), insects, and then all other animals combined. Within each grouping, animals are listed alphabetically by the name used in the text -- in most cases a common (trivial) name, where that was sufficiently diagnostic to prevent the text from becoming cluttered with Latin names. Where scientific names were used in the text, an often-used trivial name is provided here in parentheses when possible. Where common names were used in the text, the scientific names are provided here: genus with initial cap and species (both in *italics*) for names of individual species; subfamily (ending in "-inae"), family ("-idae"), order (preceded by "O."), class ("C.") or phylum ("P.") names for higher taxa. Where it could be determined that a name has been changed, the newer name is used. Very general references to birds, mammals, animals, insects, plants, invertebrates and the like are not indexed. A page number in *italics* indicates that a drawing of the species occurs on that page.

Index to Subjects

Page numbers in *script* refer to passages in which a topic or word is defined, used in a defining sense or treated extensively. Page numbers in *italics* indicate that the topic appears in a table or figure. In most cases, only one form of a key word is indexed; *e.g.*, for "melanism," "melanistic," "melanin," *etc.* see "melanism." This index also serves as a glossary for all variables and arbitrary signs and symbols used in equations, as well as for abbreviations used in the text. In most cases, only defining occurrences of such symbols are indexed; a list of equations is provided on pp. x through xii. The Greek alphabet follows the English alphabet, and all other signs and symbols are placed in arbitrary sequence at the terminus.

Σ (sigma) (summation operator)
φ (phi) (angle of diffuse re-
    flection), 64-5
ψ (psi) (magnitude of visual
    sensation), 120

= equal to
≠ not equal to

≐ approximately equal to
≡ defined as (equivalent)
> greater than
>> much greater than
≥ equal to or greater than
∝ proportionality operator, 16-7
↦ mapping operator, 17
↑ monotone operator, 17